SCIENCE AND GOVERNMENT

The Godkin Lectures at Harvard University, 1960

The Godkin Lectures on the
Essentials of Free Government and the Duties of the Citizen
were established at Harvard University
in memory of Edwin Lawrence Godkin (1831–1902).

SCIENCE AND GOVERNMENT

BY C. P. SNOW

Foreword by
Lord Robert May of Oxford

HARVARD UNIVERSITY PRESS
Cambridge, Massachusetts, and London, England
2013

Printed in the United States of America

Cataloging-in-Publication Data is available from the Library of Congress.

ISBN 978-0-674-07237-4 (pbk.)

CONTENTS

FOREWORD

As a young post-doc newly arrived at Harvard University's Department of Engineering and Applied Physics, I had the good fortune to hear Charles Percy Snow deliver his Godkin Lectures in 1960. Little did I imagine then that three and a half decades later I would be the Chief Scientific Advisor (CSA) to the UK Government and Head of its newly established Office of Science and Technology.

Snow's lectures, based on archival material as well as his own involvement at the time as a senior scientific civil servant, give a vivid account of the conflicts between Henry Tizard and F. A. Lindemann in relation to what today might be called science-based choices about resource allocation in the years leading up to, and during, World War II.

With the prospect of conflict on the horizon, in 1935 Tizard was asked by a Labour government to head a committee of experts studying how best to defend the UK, especially against bombing from Germany. He reinforced his committee with other distinguished scientists, including Patrick Blackett and Edward Appleton who would later win Nobel prizes. The committee's emphatic recommendation was for concentrating funding on what is now called radar. This, mark you, was before it was known to work!

Lindemann, an indifferent scientist but skilled social climber, had ingratiated himself with Winston Churchill, and was strongly opposed to the unproven radar. He offered instead a diversity of ineffective—not to say plain silly—alternatives. At this time Churchill had little political influence (being in opposition), and the decisions of Tizard's committee prevailed. Many believe their prescient decision played a decisive role in Britain's survival in the 1940 air battle.

By 1940, with WWII in full swing, Churchill as prime minister consistently took scientific advice only from Lindemann; the wiser and better-informed opinions of people like Tizard, Blackett, and others were simply excluded.

In 1942, a major row arose about the strategic bombing of German cities. But in Round 2 of the Tizard-Lindemann battle, the only advice seen by the Cabinet came from Lindemann. Tizard and Blackett independently examined Lindemann's statistics, showing his cost-benefit estimates in error by factors of 5 and 6, respectively; more accurate post-war studies revealed a ten-fold overestimate. In short, this costly bad decision was reached by Lindemann and Churchill—advisor and advisee—excluding all contrary evaluation by experts.

Against this background, Snow draws a variety of conclusions. It is interesting to sketch the extent to which they anticipate the UK's "Guidelines for Science Advice in Policy Making." These were prompted by the contretemps of Bovine Spongiform Encephalopathy (BSE or "Mad Cow Disease") in the UK in 1989, when an expert report recognised that the analogy with scrapie in sheep (caused by a prion broadly akin to that causing BSE, and which has no known effect on humans) suggested that BSE posed no threat to humans, but that one could not be sure. This cautiously tentative advice was rapidly transmogrified, by ministers wishing to avoid harming the beef industry, into "nothing to worry about." The relevant cabinet minister for agriculture and food memorably appeared on TV in May 1990, attempting to feed his four-year-old daughter a beefburger (she wisely refused). Sadly, the first human fatalities, unambiguously caused by BSE, followed soon after.

I was by the mid 1990s the first CSA appointed at the Mandarin, the highest, "Permanent Secretary" level in the UK Civil Service, and suggested to the conservative prime minister, John Major, that he issue the above-mentioned Guidelines. Their essence is: when there arises a problem which involves important questions at or beyond the boundaries of known science, get together an appropriate group of the best scientists in the relevant area; deliberately solicit dissenting opinions; do not exclude people with "interests," but make sure such interests are acknowledged; and—above all—do everything openly (no consensus-forming behind closed doors) and admit, indeed emphasise, uncertainties.

These UK Guidelines have gone through several itera-
tions of revision and review by now, growing bulkier at
each iteration, but with the above principles remaining
central.

Snow emphasises one major way in which politics can
work against well-judged advice. He calls it "court politics," with an individual or group exerting undue influence through "a man who possesses a concentration of
power." Lindemann's defeat of Tizard in Round 2 is a perfect example. The existence of a CSA, supported by the
Guidelines sketched above, helps combat such "court politics." This, of course, also requires the CSA to have direct
access to policy makers at the relevant level. But even beyond this, and trickier to fulfill, it ideally requires that
the CSA (or equivalent) have the cultural anthropology
skills and/or game-playing tactical skills to operate in
this arena.

Conclusions are also drawn by Snow about some of the
personal attributes a successful Science Advisor should
not have. One of these negative attributes is "euphoria of
gadgets"—an excessive enthusiasm for a particular device;
here the nearer the individual is to the physical presence
of his own gadget, the worse the judgement is going to
be. Another is the "euphoria of secrecy," which perverts
sound judgement in much the same way as the "euphoria
of gadgets." Snow refers to individuals he has known
whose judgement has "run wild [as a result of] . . . hugging to oneself a secret about one's own side or about the
other."

In concluding his lectures, Snow highlights his belief

that administrators in government are by temperament "active men," but he stresses the putative associated weakness that they "live in the short term." Here I part company with Snow. I believe the problem is not so much "short-termism," but rather that the Civil Service in general tends to mistake forming committees, consulting, and preparing reports for actually getting something done—more tersely, to mistake process for product. Be this as it may, Snow's view leads him to stress the need for such civil servants' putative focus on the immediate to be counterbalanced by the "gift of foresight" characteristic of "scientists [who] have it within them to know what a future-directed society feels like, for science itself, in its human aspect, is just that." This is powerful stuff, and indeed resonates strongly with Snow's justifiably well-known earlier publications on the Two Cultures.

Snow's world is half a century distant from today, and much has changed. However, although I would not endorse all the opinions conveyed in these Godkin Lectures, I believe they express many abiding truths about the uses and abuses of scientific advice in guiding policy decisions, often under significant uncertainty.

This remains an important book. It richly deserves its republication.

Lord Robert May of Oxford, OM AC FRS
Oxford University
September 2012

PREFACE

I should like to thank the President and Fellows of Harvard College for the honour of being asked to deliver these lectures.

Sections II–VIII (pp. 3–46) are concerned with a piece of recent history. For this, my main written sources have been the Tizard papers. As I have said in the text (p. 5), I am deeply grateful to Dr. Peter Tizard, Lady Tizard, and Mr. R. H. Tizard for the chance to study and use these sources: they are probably the richest of any in England connected with the scientific side of the 1939–45 war.

I have also had the good luck to be able to talk to many of the people who were involved in those events. The list of names would be too long to give here: but I owe a special debt to Dr. Noble Frankland, Historian of the Air Ministry and now Director of the Imperial War Museum, Dr. A. V. Hill, Professor P. M. S. Blackett, and Dr. A. P. Rowe.

During 1960 I happen to have spent some time in four of the great universities of the world: the English Cambridge, which of course I love; the Lomonosov University of Moscow; the University of California at Berkeley, which was kind enough to ask me to spend the autumn there; and Harvard. I have much feeling for all these institutions, and I do not relish praising one more than the others. And yet I felt again, as I came to Harvard for the third time, that this was in many ways the most splendid university I had ever set foot in. Giving three lectures on three successive nights is pretty rough on the lecturer, not to speak of his audience. That is what, as Godkin lecturer, I have just had to do. My impression of Harvard's splendour has survived the experience, and therefore, it seems to me, will remain with me for good.

C. P. S.
Leverett House, Cambridge
December 2, 1960

SCIENCE AND GOVERNMENT

One of the most bizarre features of any advanced industrial society in our time is that the cardinal choices have to be made by a handful of men: in secret: and, at least in legal form, by men who cannot have a firsthand knowledge of what those choices depend upon or what their results may be.

When I say "advanced industrial society" I am thinking in the first place of the three in which I am most interested—the United States, the Soviet Union, and my own country. And when I say the "cardinal choices," I mean those which determine in the crudest sense whether we live or die. For instance, the choice in England and the United States in 1940 and 1941, to go ahead with work on the fission bomb: the choice in 1945 to use that bomb when it was made: the choice in the United States and the Soviet Union, in the late forties, to make the fusion bomb: the choice, which led to a different result in the United States and the Soviet Union, about intercontinental missiles.

It is in the making of weapons of absolute destruction that you can see my central theme at its sharpest and most dramatic, or most melodramatic if you like. But the same reflections would apply to a whole assembly of decisions which are not designed to do harm. For example, some of the most important choices about a nation's physical health are made, or not made, by a handful of men, in secret, and, again in legal form, by men who normally are not able to comprehend the arguments in depth.

This phenomenon of the modern world is, as I say, bizarre. We have got used to it, just as we have got used to so many results of the lack of communication between scientists and nonscientists, or of the increasing difficulty of the languages of science itself. Yet I think the phenomenon is worth examining. A good deal of the future may spring from it.

In the West, we have not been very good at looking at this singularity with fresh and candid eyes. We are too apt to delude ourselves with phrases like "the free world," or "the freedom of science." None of those phrases is meaningful when we are concerned with the kind of choice I am describing. Such phrases only obscure the truth. I shall come back to that point later. For the moment I will just say that all societies, whatever their political structure or legalistic formulations, are going to be faced with this same type of choice so long as we have nation-states, and that the results are going to be not only significant, but much too significant.

I know that we can draw diagrams of political responsibility which are able to make us feel that everything can be reconciled with the principles of parliamentary government. But if we do, we shall not even begin to understand what is really happening. We shall fool ourselves, as we do too often, with that particular brand of complacency, of lack of gravity, which is one of the liabilities of the West, growing upon us perhaps as we become more affluent.

The first thing, it seems to me, is to try to understand what really happens. "We must learn to think," Don K. Price has written, "without making use of the patterns or models taken for granted by most of the text books."[1] It is harder than it sounds.

No one who has ever thought at all about the relations of science and government, much less anyone who has experienced part of them directly, is likely to think that positive conclusions are going to be either firm or easy to come by. Most of the concepts that administrative theorists use are at best rationalisations, not guides to further thought; as a rule they are unrealistically remote from the workaday experience.

No one that I have read has found the right answers. Very few have even asked the right questions. The best I can do is tell a story. The story is intended to contain a little of something which actually did happen. I shall not pretend that the story is not supposed to bear some relation to our present problems. I shall try to extract a few generalisations from it, or, to be more sensible, a few working rules.

II.

This story is about two men and two choices. The first of the two men is Sir Henry Tizard. Let me declare my interest straight away, as they say in English board rooms. I believe, along with a number of Englishmen who are interested in recent military-scientific history, that Tizard's was the best scientific mind that in England has ever applied itself to war. I further believe, although in general I take a pretty Tolstoyan view of the influence of distinguished men upon events, that of all the people who had a share in England's surviving the air battles of July to September 1940, Tizard made a contribution at least as great as any. It has not yet been properly recognised. As he himself wrote in his diary on May 8th, 1945, when he was living in what for him was high-level exile, as president of Magdalen College, Oxford, "I wonder if the part that scientists have played will ever be faithfully and fully recorded. Probably not."[2]

To an American audience, it is natural that I should have to introduce him from scratch: but if I was speaking of him to most English audiences, I should have to do the same. In fact I have never spoken of him before, and I am very glad that I should do so for the first time in the United States. He had much feeling for America and American science. It was owing to him, as we shall see, that, sixteen months before the United States came into the war, American scientists were told all that the English were doing and all they knew. That gesture of bold trust, forced through by him, and very like his temperament, saved both our countries quite an appreciable bit of time in the Hitler war.

I happen to know that he would have liked me to talk

about him, because I once threatened him with it. He said: "At least I can trust you to do it with the gloves off." He meant, of course, as he said himself when writing of Rutherford, that with characters big enough one ought not to be polite. His family are also sure that he would have relished being treated so, and I have been given unqualified access to the Tizard papers. He wrote quite a lot about himself. He began an autobiography and he kept a number of fragmentary diaries. Towards the end of his life, like a good many men who have played a part in history, he wanted his own end of the record to be kept straight. Although I knew him well, I have drawn on this documentary material as well as on other written sources. There is very little in what follows which is my own opinion or unsupported impression. When there is, I shall try to make it clear.

What was he like? Physically he did not alter much from middle age, when I first met him, until he died in 1959. He was English of the English. His whole appearance, build, and manner were something one does not often see outside England, or even outside the English professional class from which he sprang. He was not pretty. There were times when he looked like a highly intelligent and sensitive frog. His hair, what was left of it, was reddish. His face was unusually wide across the jaw line. But his expression was transfigured by his eyes, which were transparent light blue, sparkling with dash and interest. He was middle sized, and like nearly all successful men of affairs, he was in a muscular sense strong.

But that tough physique, that alert, confident, command-ing manner, that warm rasp of a voice, hid certain dishar-monies. He was not all of a piece.

He came into a room, and he had an authority, a pug-nacity, that made men attend to him. He had a lively sa-tirical tongue, of a kind that seemed a little stylised to my generation. "Andrade [who was looking after wartime in-ventions] is like an inverted Micawber, waiting for some-thing to turn down." Of the personal antagonism with which I shall soon be dealing: "The hatchet is buried for the present: but the handle is conveniently near the sur-face." And so on. There were heaps of Tizardisms—but they were to an extent misleading.

True, he knew he was a gifted man; he knew his own capacity pretty well; but the confidence which made men follow him was not the deep-rooted, relaxed confidence of those who have their creative achievement safely behind them—the relaxed creative confidence, for example, of his idol Rutherford. Tizard did not always find himself easy to live with. The bold face he put on did not completely mask the strains of his inner life.

In the same way, his tough powerful physique was not as impregnable as it looked. All his life he seems to have been vulnerable to infections, suddenly knocked out by mysterious high temperatures. He was lucky in his family, and had sons of very high ability: but he had a great need for affection, not only in his family, but among his friends. Friendship mattered more to him than it would have done if he had been the self-sufficient man he looked. Fortunately for him, he had the energy and warmth to make friends of all ages. I sometimes thought he was at his happiest in the Athenaeum—he had the curious distinction of being able to make the Athenaeum cosy—among people who not only admired him, but were fond of him.

He was born in 1885. His father was a naval officer—a naval officer of strong scientific leanings, who became assistant hydrographer to the Navy and a Fellow of the Royal Society, but first and foremost a naval officer. That had a direct importance to Tizard, both in his attitudes and in what he was able to achieve. All his life he had the simple, unquestioning, absolute patriotism of a regular officer: and he had a complete intuitive understanding of what soldiers and sailors were like. Except for a physical chance, in fact, he would have been one himself. He would—as a matter of course in such a family—have entered the Navy, if it had not been discovered, just before the examination, that he had a blind patch in one eye. Tizard says, "I must have taken this verdict philosophically at the time, for I don't remember being disappointed or relieved: but it was a bad blow for my father . . . He went to a friend in the Admiralty and said, 'What would you do with a boy who cannot get into the Navy?'"[3]

These traditional loyalties were very deep in Tizard. In scientific and technical things his mind was radical: but emotionally he remained until he died bound to that upright, intelligent, dutiful, conservative line. His family were always short of money. Running true to form of the conservative English service families, they both had a certain contempt for money and were constantly worried about it. That stayed so with Tizard. He was worried about money till his death. He never made any, and when he retired from the public service no proper provision was made for him, owing to the changes and chances of his career. His one bitter complaint, in his old age, was that he did not know how he was going to live.

Instead of entering the Navy, he went through an orthodox professional English education—Westminster and Oxford. He was dazzlingly clever at anything he put his hand to. Later on he thought he might have made a goodish academic mathematician, and wished he had tried. Actually, he specialised in chemistry, which was at the time the only adequate scientific school in Oxford. Oxford is now, of course, highly developed in scientific subjects, and it is a bit startling to be reminded that the young Tizard in 1908, bursting with both academic honours and promise, anxious to make a start in research, could find no one in Oxford to work under. Like other bright young Englishmen and Americans of that period, he decided that Germany was the place to find the masters of research. He went off to Berlin to work under Nernst.

As it turned out, he did not bring off anything of scientific interest during his year there. But he brought off something else. For it was in Nernst's laboratory that he first met the other main character in this story. There is a difficulty about this other character because of the English habit of changing names and styles. Thirty odd years later, as the right-hand man and grey eminence of Winston Churchill, he became known as Lord Cherwell. But nearly all the way through his friendship and enmity with Tizard he was called F. A. Lindemann. That is the name by which Tizard in his papers always refers to him. For clarity's sake I shall stick to the same convention.

III.

These two young men met in Berlin in the autumn of 1908. We do not know the exact circumstances. It would be nice to know, for even if we eliminate what was to happen, they were two of the most remarkable young men alive, and there cannot have been many such meetings. Lindemann was, by any standards, a very odd and a very gifted man, a genuine heavyweight of personality. I did not know him as well as I did Tizard, but I talked to him a good many times. As he thought I was relatively sensible about the job I was doing, he gave me some tough support. He even made a speech about me in the House of Lords.[4] More important than that, as far as I was concerned, his was the sort of character that makes a novelist's fingers itch. So, although in the two issues I am going to use for analytical purposes I have no doubt that he was wrong and Tizard right, I have a soft spot for him and a complex of respect. I do not think that I should be so interested in the Tizard-Lindemann struggles if I did not have that kind of feeling for both men.

I said that Tizard was English of the English. Lindemann was quite un-English. If one met him for the first time in middle age, I have always thought that one would have taken him for a Central European business man —pallid, heavy featured, correctly dressed, one who had been a notable tennis player in his youth and was now putting on weight. He spoke German as well as he did English, and there was a faint Teutonic undertone to his English, to his inaudible, constricted mumble. No one seems to know to this day what his father's nationality was.[5] He may have been a German or an Alsatian. It is possible, though I doubt it, that he was Jewish. No doubt this rather silly mystery will be cleared up in the official biography which Lord Birkenhead is now writing. But it is certain that Lindemann's father was distinctly rich, and Lindemann himself, unlike Tizard, had the attitude to money of a rich man, not of a member of the professional Establishment.

There was a similar sharp difference in the nature of their patriotism. As I have said, Tizard's was the patriotism of a naval officer, which came to him as naturally and unselfconsciously as breathing. Lindemann, who was not an Englishman but became one, had the fanatical patriotism of someone who adopts a country which is nevertheless not, in the deepest sense, his own. No one cared more about England than Lindemann, in his own way: but it was a way that, with its flavour of the patriotism of the converted exile, struck men like Tizard as uncomfortable and strained.

A great deal else of Lindemann's personality struck them also as uncomfortable and strained. About him there hung an air of indefinable malaise—so that, if one was drawn to him at all, one wanted to alleviate it. He was formidable, he was savage, he had a suspicious malevo-lent sadistic turn of what he would have called humour, though it was not really that. But he did not seem, when it came to the most fundamental things, to understand his own life, and despite his intelligence and will, he did not seem good at grappling with it. He enjoyed none of the sensual pleasures. He never drank. He was an extreme and cranky vegetarian, who lived largely on the whites of eggs, Port Salut cheese, and olive oil. So far as is known, he had no sexual relations. And yet he was a man of in-tense emotions.

Tizard, whose emotions were also deep and difficult to control, had an outgoing nature, which, luckily for him, found him wife and family and friends. Lindemann's pas-sions were repressed and turned in upon himself. You could hear the difference in their kind of joke. Tizard, as I mentioned, had a tongue which was harsh, which could be rough with pretentious persons, but which was in the long run good-natured. Lindemann's had the bitter edge of repression.

I remember being in Oxford one morning when the Honours List had been published. I think this must have been during the war. I was talking to Lindemann. I hap-pened to remark that the English honours system must cause far more pain than pleasure: that every January and June the pleasure to those who got awards was nothing like so great as the pain of those who did not. Miracu-lously Lindemann's sombre, heavy face lit up. His brown eyes were usually sad, but now they were glowing. With a gleeful sneer he said: "Of course it is. It wouldn't be any use getting an award if one didn't think of all the people who were miserable because they hadn't managed it."

In that kind of venom, in almost everything he did, he was much more intense than most men. His passions were a bit bigger than life-size; they often took on the inflated monomania of the passions in Balzac's novels. He was altogether a bit bigger than life-size. As I have already said, he was a character who made a novelist's fingers itch. And yet, thinking of him and Tizard, I am not sure which would interest me more as a novelist. When I was younger, Lindemann certainly. Now that I have found my interest gradually change from what we call "abnormal" to "normal" personalities—I am using these words, of course, as a shorthand jargon—I think it might be Tizard. He was externally a far less odd man than Lindemann. In the structure of his personality he was probably more complex.

IV.

One would like to know what they talked about, in Berlin that winter of 1908. Science, of course. Both had an unshakable faith that science was the supreme intellectual manifestation of the mind of man, a faith they never lost. Tizard had strong interests in literature, but Lindemann none, nor in any other art. Maybe they talked about politics. Both were conservative, but Tizard had the receptive tolerant conservatism of the Establishment, while Lindemann was eccentrically, and often extremely, reactionary.

I do not think they talked of love or young women, as
men of that age might be expected to.

There was a romantic story, dear to some in Whitehall
who met them in the days of their power and un-patch-
up-able quarrels, that they had once been inseparable. I
believe, from the quotation from Tizard's autobiography
which follows and from other evidence, that that is over-
doing it. It is true that Tizard was writing long after the
event: but he was also deliberately composing his autobiography with the Lindemann feud as its chief dramatic
conflict, and he was too much of a natural storyteller to
have underplayed their original friendship, if honesty had
not compelled him to do so.

F. A. Lindemann and I became close but not intimate
friends. [This is the first reference to Lindemann in
the autobiography.] There was always something about
him which prevented intimacy. He was one of the cleverest men I have known. He had been to school in
Germany, and talked German very well—as well as he
talked English—and was fluent in French. He was a
very good experimenter. He also played games well. He
wanted me to share rooms with him [in Berlin] but I
refused. I think my chief reasons for doing so at the
time were that he was much better off than I was and I
could hardly compete with his standard of living, and
also that we should be speaking English all the time,
for he would take no trouble to teach me German. It
was lucky that I refused because we had a minor row
later on. I had discovered a gymnasium in Berlin which

was run by an ex-lightweight champion boxer of England, so I used to go there for exercise. I persuaded Lindemann to join and box with me.

Now one of his greatest defects was that he hated anyone of his own age to excel him in anything. He was a clumsy and inexperienced boxer, and when he found that I, who was much shorter and lighter than he was, was much quicker with my hands and on my feet, he lost his temper completely, so much so that I refused to box with him again. I don't think he ever forgave me for that. Still, we remained close friends for over twenty-five years, but after 1936 he became a bitter enemy.[6]

After that year in Berlin, Lindemann stayed in Germany, where he had his entire education, high school, undergraduate, and postgraduate. Tizard returned to England and became a scientific don at Oxford. As he wrote himself,[7] in view of his subsequent career it was strange that he did not remember taking the slightest interest in the application of science to war before 1914. At that time, all his ambitions were in pure science, and they were broken only by the beginning of the war and by a friendship, a hero-worshipping friendship, with Rutherford. That sounds a paradox, since Rutherford was the supreme creative expression of pure science, but it makes good psychological sense, and I will deal with it in a moment.

In the 1914-18 war both Tizard and Lindemann, in their early thirties, played picturesque parts. Both happened to be not only brave, but abnormally brave, in the starkest physical sense. Both happened to find their way into the primitive aircraft experimentation of the time. They volunteered for it, because they were not allowed to fight behind machine guns. Tizard was offered flying training, but only in weather *too rough* for the normal flying cadets. "Done," he said. Lindemann, for experimental purposes, deliberately put his aircraft into a spinning nosedive. It was against the statistical probabilities that either remained alive, let alone both.

After the war their lives interweaved again. Tizard went back to teach chemistry at Oxford. He put in a word with the electors to the chair of experimental philosophy on behalf of Lindemann,[8] who was duly elected, much to the astonishment of the English physicists, since Lindemann had never been inside an English university. Lindemann became godfather to one of Tizard's children. For two or three years it seemed that they might lead a scientific renaissance in Oxford, the first since the seventeenth century.

But then something began to happen to them both—quite clearly to Tizard, more foggily to Lindemann, who had far less introspective insight. What happened was simple. They knew they were never going, by high standards, to make a success of pure science. Tizard was explicit about it, both in conversation, "I knew I should never be any *real* good," and in his autobiography, "I now convinced myself that I would never be outstanding as a pure scientist. Younger men were coming on of greater ability in that respect."[9] By this he meant that he could not fight at the same weight as Rutherford and his friends. Rutherford, who had become a major influence in his life, had set him a standard to judge scientific achievement by. Tizard did not expect to be a Rutherford. They occurred once in three hundred years. But he was a proud man, he had a sense of his own powers, and he wanted at least to be as good as the next rung down. He felt he was not, and that settled it.

In all this I am reminded of Alfred Kazin's comment about Englishmen weighing themselves and each other up as though they were so much horseflesh. All I can say is, it happened. With Lindemann, it took more time, and it was not so incisive. But he was an even prouder man than Tizard, and internally more convinced that he had a great intellect. He could not tolerate not being able to compete on the one hand with Rutherford and the new generation of Rutherford's pupils, Chadwick, Cockcroft, Kapitza, Blackett, or, on the side of mathematical physics, with Bohr, Heisenberg, Dirac, and a dozen others. It just wasn't good enough. So they each, one consciously and the other gropingly, took their separate ways out.

It is interesting to wonder whether they were right. If they had had more creative confidence, which they both seriously lacked, would they have left a real scientific memorial behind them? After all, they were out of comparison more intelligent than many scientists who have made major discoveries. In his last years Tizard certainly—here I cannot speak for Lindemann—would have given up all his other achievements if he could have had even a quarter of a Rutherfordian oeuvre to his credit. With more luck, with less pride, could he, could either of them, have done it? As I think that, I hear, from twenty years ago, the clear voice of G. H. Hardy: "For anything worth doing [by which Hardy meant creative work, which he took for granted was the only thing worth doing] intelligence is a very minor gift."

Probably, one is forced to believe, their intelligence would not have compensated, and they were right when they contracted out. Tizard had a very broad scientific comprehension. He was the kind of scientist, of which Willard Gibbs was a supreme example, who builds great systems: but Tizard had not the special insight which would have let him see which system, in his own time, was there to be built. Lindemann was the opposite. Apart from his zest in destructive criticism, he was a gadgety scientist, inventive, on the lookout for ingenious tricks. To make use of that gadgety talent, one has to have the obsessive force that can keep one thinking over one device for year after year. Aston could do that, so could C. T. R. Wilson, so could Thomas Merton.[10] But Lindemann soon got tired. That was why he remained an amateur among professionals: which, by the way, was how the leaders of physics, such as Rutherford, always regarded him.

V.

So, though they both became fellows of the Royal Society at an early age—earlier than they could have hoped to become in the conditions of today—Tizard and Lindemann slipped out of pure science. And their ways of slipping out brought about the two great collisions. Tizard became a high-level scientific administrator. That was less than forty years ago, but England had only just begun to spend money on applied science. It was during the 1914–18 war that the Department of Scientific and Industrial Research was started. Tizard, who had made a great reputation in applied science during the war,[11] succeeded to the job of permanent secretary, that is, the chief official responsible to a minister. Such chief officials in England have greater power, and more influence in determining policy, than their opposite numbers in the United States. In England they are right at the heart of the Establishment, and in a good many ways are more steadily and continuously important than their political bosses. Tizard fitted into that world from the start. He was not exactly an administrator's administrator, but he was liked and trusted by the high officials. They were in origin and in general attitude, if we forget his streak of scientific radicalism, very much like himself. He liked Whitehall. He liked the corridors of power. He liked the Athenaeum. He liked his colleagues, men like himself devoted, upright, and tough, though nothing like so outspoken as he was. When he moved off to become rector of the Imperial College, London, in 1929, he did not leave this inner English official world.

During those same years, Lindemann was making his way in quite a different English world—the world of high society and conservative politics, which at the time, when

"Society" had a practical function that is now obsolescent or dead, still overlapped. It may seem odd that it was so easy for someone without any social connections, who was not even English by birth, who was about as little like a typical specimen of the English upper classes as one can comfortably imagine, to penetrate right into the inner sanctums. But it is really very simple. It is only a puzzle if one approaches English society with Proustian illusions. Lindemann was rich: he was also determined. For generations English society has been wide open to, defenceless against, rich and determined men. The more so if they happen to be intelligent. So within a matter of months rather than years Lindemann was eating his singular vegetarian meals at a good many of the great English houses. He became known among smart people, with somewhat unfortunate infantilism, as "the Prof." He was very soon an intimate of Lord Birkenhead (F. E. Smith), and through Birkenhead he met Winston Churchill and began, apparently almost at first sight, a friendship which determined the rest of his life.

This friendship was utterly loyal on both sides, and continued so until Lindemann's death. A good deal of Lindemann's social progress was snobbish, an escape from inner defeats. But his devotion to Churchill was the purest thing in his life. It was quite unaffected, or perhaps more strengthened than weakened, by Churchill's ten years out of office (1929-1939) when it looked as though he were one of a hundred great men *manqués,* one of those with a brilliant future behind them. Churchill's loyalty to Lindemann was also absolute. Later on, Lindemann, as Churchill knew well enough, became a cause of friction with Churchill's other intimates, something of a political liability, Churchill didn't budge an inch.

Why this friendship? a good many people have asked. They appeared a pretty incompatible pair. Churchill does not seem at first glance the obvious soulmate for a fanatical ascetic, a teetotal nonsmoking vegetarian. But the question, like a similar question about Roosevelt and Harry Hopkins, is without meaning unless one knew both men, not just well, but as well as they knew each other. Why any friendship, as far as that goes?[12]

VI.

In 1934 both Tizard and Lindemann were nearly fifty. Of the two, Tizard had been by a long way the more successful, though even he, judged by the standard he set himself, had not lived up to his promise. He was a trusted man of affairs, he had been knighted, he was head of a university institution, but in his own eyes he had not done much.

As for Lindemann, he had done much less. The professional physicists did not take him seriously as a scientist, and dismissed him as a cranky society pet. Scientifically his name was worth little. He was the intimate friend of a politician whose name was scarcely worth as much.

Then, quite suddenly, Tizard was given the chance for which he was made. England was strategically in a desperately vulnerable position, for reasons—the tiny size of the country, the density of the population—which apply more harshly today. In 1934 Baldwin was the main figure in the government, and it was only two years since he had said lugubriously: "The bomber will always get through."

In public, rebellious politicians like Churchill were attacking the whole of the government's defence policy. In secret, the government scientists, the military staffs, the high officials, were beating round for some sort of defence. There was nothing accidental about this. It was predictable that England, more vulnerable to air attack than any major country, would spend more effort trying to keep bombers off. But there was something accidental and unpredictable in Tizard being given his head.

The Air Ministry, under the influence of their scientific adviser, H. E. Wimperis, himself prodded by a bright young government scientist called A. P. Rowe,[13] set up a Committee for the Scientific Study of Air Defence. Its terms of reference were as flat as usual: "To consider how far advances in scientific and technical knowledge can be used to strengthen the present methods of defence against hostile aircraft." The committee was nothing very important to start with. No one took much notice when its membership was announced. There may have been slight curiosity about the appointment, which was entirely due to Wimperis,[14] of Tizard as chairman. The appointment would not and could not have happened, though, if Tizard had not been so well connected in official life.

Well, that committee was called the Tizard Committee almost from its first meeting. It is slightly touching that in his diary Tizard, who could not use that title, never seems to have been quite certain what its official title really was.

From the first meeting on January 28th, 1935 he gripped the problems. This was the job for which he was born. Quite soon, by the summer of that year, small ripples of confidence oozed under the secret doors and penetrated Whitehall, almost the only ripples of confidence that touched the official world during those years. Tizard insisted on a very small committee which he chose himself. Wimperis had to be there, Rowe was brought in as secretary, but at the beginning there were only two members of independent standing, A. V. Hill and P. M. S. Blackett. Both of these were eminent scientists, of a quite different order of accomplishment from Tizard or Lindemann. Hill was one of the most distinguished physiologists in the world and had won a Nobel prize in 1922. Blackett, who was only thirty-seven at this time, was one of Rutherford's most brilliant pupils, and later himself won a Nobel prize.[15]

I doubt if their scientific stature was Tizard's first reason for choosing them. He was an exceptionally good picker of men. Like all good pickers, he was not distracted by much; he was thinking of what the men could do. It did not matter to him that Hill was a very unorthodox conservative, hotly out of sympathy with the Baldwin-Chamberlain policy, the policy of Tizard's own Establishment friends. It did not matter to him—as it would certainly have done to more cowardly men—that Blackett was a radical, the most distinguished figure among all the radical young scientists, who were bitterly antifascist and who distrusted every move that our own government made. I can say that without hedging, because I was one of them myself.

Tizard did not care. He knew that Hill and Blackett were men who were equipped not only with technical insight, but with strong characters and capacity for decision. That was what he wanted. There was not much time to play with. And I have, though I can produce no evidence for it, a strong feeling that he wanted just one other thing. He wanted the members of his committee to have a natural sympathy for and identification with military men. Hill had been successful in the Army in the first world war, and had edited a classical work on antiaircraft gunnery. Blackett, before he turned to physics, had been a professional naval officer.

That was a factor in their success, I am convinced. Because the first task was not only a scientific choice, which they made quickly, but also an effort of indoctrination in the services (and a mutual give and take between serving officers and scientists) without which the choice was useless. The choice itself faced them like an "either/or." *Either* what was later called by its American name of radar, but in these aboriginal days was known as R.D.F., was the device to back: *or* there was nothing to back.

The committee made up its mind about that before the device really existed. Watson Watt, who was the pioneer of radar in England, working in the Radio Research Laboratory of the D.S.I.R., had done some preliminary experiments. This device might, not certainly but possibly, work in real war in three or four years. Nothing else possibly could. Tizard, Hill, Blackett had faith in their own reasoning. Without fuss, and without backward glances, the choice was made. That was only a resolution on paper, and they had to make it actual.

The administrative mechanism by which this was done is itself interesting. In form the Air Minister, Lord Swinton,[16] arranged for a new high-level committee which was to act as a subcommittee of the Committee of Imperial Defence. Over this new body he himself presided, and on to it was brought the government's chief military critic, Winston Churchill. In fact, however, one has got to imagine a great deal of that apparently casual to-ing and froing by which high English business gets done. As soon as the Tizard Committee thought there was something in radar, one can take it that Tizard would lunch with Hankey[17] at the Athenaeum; Hankey, the secretary of the Cabinet, would find it convenient to have a cup of tea with Swinton and Baldwin. If the Establishment had not trusted Tizard as one of their own, there might have been a waste of months or years. In fact, everything went through with the smoothness, the lack of friction, and the effortless speed which can only happen in England when the Establishment is behind one. Within a very short time the Tizard Committee were asking for millions of pounds, and getting it without a blink of an eye. Two successive secretaries of the Cabinet, Hankey and Bridges,[18] did much more than their official duty in pushing the project through.

The second active job was, in particular, to persuade

the serving officers of the Air Staff that radar was their one hope and, in general, to make scientists and military people understand each other. Here again this might have been impossible. In fact, with the exception of those concerned with bombing policy, the senior officers were ready to be convinced as soon as Tizard started to talk.[19] They often thought of putting him in uniform: but that would have defeated his whole virtue as an interpreter between the two sides. "I utterly refuse to wear a busby," he used to say. Fairly soon he had not only got radar stations in principle accepted and hoped for, but also succeeded, with the help of Blackett's exceptional drive and insight, in beginning to teach one lesson each to the scientists and the military, lessons that Tizard and Blackett went on teaching for twenty years.

The lesson to the military was that you cannot run wars on gusts of emotion. You have to think scientifically about your own operations. This was the start of operational research,[20] the development of which was Blackett's major personal feat in the 1939–45 war.[21] The lesson to the scientists was that the prerequisite of sound military advice is that the giver must convince himself that, if he were responsible for action, he would himself act so. It is a difficult lesson to learn. If it were learnt, the number of theoretical treatises on the future of war would be drastically reduced.

The committee met for the first time, as I said, in January 1935. By the end of 1935 its important decisions were in effect taken. By the end of 1936 most of those decisions

were translated into action. It was one of the most effective small committees in history. But before it clinched its
choices, there was a most picturesque row.

The committee had been set up, as we saw, from inside
the Air Ministry. One of the reasons was, no doubt, to
forestall criticism from outside, which came most loudly
and effectively from Churchill. In 1934 he had publicly
challenged the government's underestimate of the size of
Hitler's air force. His figures, which had been produced
by Lindemann, were much nearer the truth than the government's. Thus, simultaneously, there were going on the
secret deliberations and discussions of the Tizard Committee, and an acrimonious military argument in full
light in the House of Commons and the press, with
Churchill the antigovernment spokesman.

It is one of the classical cases of "closed" politics coexisting with "open" politics. Passing from one to the other,
an observer would not have known that he was dealing
with the same set of facts. By the middle of 1935 Baldwin,
who had just in form as well as fact become Prime Minister, wanted to reduce the temperature of the "open" military argument. He used the orthodox manoeuvre of asking Churchill in. Not into the Cabinet: the personal rifts
were too deep for that, but onto the new Swinton Committee, the *political* committee to which I have just referred, which was to keep a supervisory eye on air defence.

The history is very tangled at this point. No minutes
have ever been published, but if I know Hankey and his
colleagues at all—and I had the good luck to work under

them a short time later—I have not much doubt that on the one hand they felt confident that they could give the Tizard Committee its head (Tizard sat himself on the political committee and made his requests for money to it), and that on the other hand it could not do harm, and might do good, if Churchill were given exact information of what was actually being done, rather than inexact.

Roughly that was what happened, but there were other consequences. Churchill entered the political committee, retaining the right to criticise in public and insisting that Lindemann, as his personal scientific adviser, be given a place on the Tizard Committee. Both these conditions were reasonable enough: but then the private war began.

Almost from the moment that Lindemann took his seat in the committee room, the meetings did not know half an hour's harmony or work undisturbed. I must say, as one with a taste for certain aspects of human behaviour, I should have dearly liked to be there. The faces themselves would have been a nice picture. Lindemann, Hill, and Blackett were all very tall men of distinguished physical presence—Blackett sculptured and handsome, Hill ruddy and English, Lindemann pallid, heavy, Central European. Blackett and Hill would be dressed casually, like academics. Tizard and Lindemann, who were both conventional in such things, would be wearing black coats and striped trousers, and both would come to the meetings in bowler hats. At the table Blackett and Hill, neither of them specially patient men nor overfond of listening to nonsense, sat with incredulity through diatribes by Lindemann, scornful, contemptuous, barely audible, directed against any decision that Tizard had made, was making, or ever would make. Tizard sat it out for some time. He could be irritable, but he had great resources of temperament, and he knew that this was too serious a time to let the irritability flash. He also knew, from the first speech that Lindemann made in committee, that the friendship of years was smashed.

There must have been hidden resentments and rancours, which we are now never likely to know and which had been latent long before this. No doubt Lindemann, who was a passionate man, with the canalised passion of the repressed, felt that he ought to have been doing Tizard's job. No doubt he felt, because no one ever had more absolute belief in his own conclusions, that he would have done Tizard's job much better, and that his specifics for air defence were the right ones, and the only right ones. No doubt he felt, with his fanatical patriotism, that Tizard and his accomplices, these Blacketts, these Hills, were a menace to the country and ought to be swept away.

It may have been—there are some who were close to these events who have told me so—that all his judgments at these meetings were due to his hatred of Tizard, which had burst out as uncontrollably as love. That is, whatever Tizard wanted and supported, Lindemann would have felt unshakably was certain to be wrong and would have opposed. The other view is that Lindemann's scientific, as well as his emotional, temperament came in: it was not only hatred for Tizard, it was also his habit of getting self-blindingly attached to his own gadgety ideas that led him on. Whatever the motive was, he kept making his case to the committee in his own characteristic tone of grinding certainty. It was an unjustifiable case.

The issue in principle was very simple. Radar was not yet proved to work: but Tizard and the others, as I have said, were certain that it was the only hope. None of them was committed to any special gadget. That was not the cast of their minds. There was only a limited amount of time, of people, of resources. Therefore the first priority must be given to radar—not only to making the equipment, but to making arrangements, well in advance even of the first tests, for its operational use. (It was in fact in the operational use of radar, rather than in the equipment, that England got a slight tactical lead.)

Lindemann would not have any of this. Radar was not proved. He demanded that it should be put much lower on the priority list and research on other devices given the highest priority. He had two pet devices of his own. One was the use of infra-red detection. This seemed wildly impracticable then, to any of the others and to anyone who heard the idea. It seems even more wildly impracticable now. The other putative device was the dropping, in front of hostile aircraft, of parachute bombs and parachute mines. Mines in various forms had a singular fascination for Lindemann. You will find Rube Goldberg-like inspirations about them—aerial mines, fluvial mines, and so on—all over the Churchillian minutes from 1939–1942.[22] They keep coming in as a final irritation to a hard-pressed man in Tizard's records of his conversations with Churchill. All these mine inspirations originated from

Lindemann. None of them was ever any practical good at all.

For twelve months Lindemann ground on with his feud on the committee. He was tireless. He was ready at each meeting to begin again from the beginning. He was quite unsoftened, quite impregnable to doubt. Only a very unusual man, and one of abnormal emotional resistance and energy, could sit with men so able and not be affected in the slightest regard.

They themselves were not affected so far as choice was concerned. Tizard went ahead with the radar decisions and they let Lindemann register his disagreement. But gradually they got worn down. Neither Blackett nor Hill was phlegmatic enough to endure this monomaniac tension for ever. In July 1936,[23] when the committee were preparing a report, Lindemann abused Tizard in his usual form, over the invariable issue of too much priority for radar, but in terms so savage that the secretaries had to be sent out of the room.[24]

At that point Blackett and Hill had had enough of it. They resigned and did not try to give an emollient excuse for doing so. Whether this was done after discussion with Tizard is not clear. No discussion was really necessary. They all believed that this friction was doing too much harm. They were all experienced enough to know that, with Churchill still out of office, they could make their own terms.

Within a short time the committee was reappointed. Tizard was still chairman, Blackett and Hill were still members. Lindemann, however, was not. He was replaced by E. V. Appleton, the greatest living English expert on the propagation of radio waves. Radar itself was an application of Appleton's fundamental work. The announcement of his name meant, in the taciturn eloquence of official statements, a clear victory for radar and for Tizard. The radar stations and the radar Organisation were ready, not perfect but working, in time for the Battle of Britain. This had a major, and perhaps a decisive, effect.

This cautionary story of the first Lindemann-Tizard collision seems to me to contain a number of lessons, some of them not obvious. But there is one, at the same time so obvious and so ironic that I shall mention it now. It is simply that the results of closed politics can run precisely contrary to the results of open politics. That is an occupational feature of the way in which closed politics works and the way in which secret choices are made. Probably not more than a hundred people had any information whatever about Tizard's first radar decision; not more than twenty people took any effective part in it, and at the point of choice not more than five or six.

While that was going on, so also was violent open politics, the open politics of the thirties, the most ferocious and deeply felt open politics of my lifetime. Nearly everyone I knew of my own age who was politically committed, that is, who had decided that fascism had at all costs to be stopped, wanted Churchill brought into the government. Partly for his own gifts, partly as a symbol of a country which was not going to let the Nazis win by default. We signed collective letters about Churchill; we used what influence we had, which in those years was not much. We wanted a government which would resist, the kind of government we finally got in 1940. That was the position, I think, of Blackett and most of my liberal friends. It was certainly my own. Looking back, I think we were right, and if put back in those years again I should do what I did then.

The ifs of history are not very profitable—but if Churchill had been brought back to office, if open politics had gone the way my friends and I clamoured and implored that it should? We should, without any question, have been morally better prepared for war when it came. We should have been better prepared in the amount of war material. But, studying the story I have just told, I find it hard to resist the possibility that, in some essential technical respects, we might have been worse prepared. If Churchill had come into office, Lindemann would have come with him, as happened later. It is then very hard to imagine Lindemann not getting charge of the Tizard Committee. As I have said, I take a pretty Tolstoyan view of history in the large. In a broad sense I cannot easily accept that these small personal accidents could affect major destinies. And yet . . . without getting the radar in time we should not have stood a good chance in the war that finally arrived. With Lindemann instead of Tizard, it seems at least likely that different technical choices would have been made. If that had been so, I still cannot for the life of me see how the radar system would have been ready in time.

These retrospective fears are not profitable. But I do not know of a clearer case where open and closed politics appear to tell such different stories and point to such different fates.

VII.

The first round in the Tizard-Lindemann duel thus went to Tizard. When war came, he had got his air defence system working. He himself became scientific adviser to the Air Ministry, and his diary between September 1939 and May 1940 is quick, hurried, and lively, written at night after visits to airfields, on the job that he did better than anyone in any country, getting scientific methods into the heads of the young officers, infusing them with his own enthusiasm and his own sense of scientific fact.

Things were going pretty well scientifically that winter, but he had another preoccupation. He had arranged for A. V. Hill to be sent on a mission to Washington, and both of them had become convinced that there were overwhelming arguments for telling the American scientists the whole of our radar and other military scientific secrets. Nearly all the English scientists agreed—Cockcroft, Oliphant, Blackett all pressed the matter. Nearly everyone else disagreed.[25] The written record is simultaneously comic and dreary, with just the kind of comic dreariness one always meets when people get seized by the euphoria of secrecy. Various nodding heads said that United States security could in no circumstances be trusted. Various others, including some who should have known better, thought the United States had nothing to offer.

Tizard became distinctly irascible, but otherwise was getting a good deal of his own way. Churchill had become First Lord of the Admiralty as soon as war broke out, and Lindemann was in Whitehall as his personal adviser. But for the moment there was an uneasy balance of power; Lindemann could not touch the air arrangements. From the papers it looks as though Tizard was as happy and as occupied in those months as at any time in his public life.

Then came May 10th, the German attack on France, Churchill in power. Tizard knew the military dangers as well as anyone alive. He also probably knew that his own days of authority would not last long. If so, his diary entries for that day and May 11th are among the masterpieces of English phlegm.

Friday, May 10. Left Oxford 9 A.M. for Farnborough by air. Saw de Burgh and discussed with him experimental work on A. I. In particular some work on frequency modulation. R. A. E. have made progress in aerial design to eliminate some of the effects of ground reflection, and Mitchell is optimistic: too much so, I think. No clear evidence that method of frequency modulation is better than the pulse method.

Saturday, May 11. From Hill Head to Tangmere. Discussed flying trials of A. I. Was told that ordinary C. H. interception was so bad that there was little hope of

getting good A. I. interception by night until day interception was improved. I told them that I thought it
better to concentrate on day interception with the help
of A. I. rather than do night interceptions now.[26]

The German armies cut through France. Churchill and
Lindemann were in 10 Downing Street, getting ready to
take control of the war, including the scientific war. Tizard's diary goes on just like those two extracts, full of his
actions, advice, memoranda. Of course, there is a great
inertia behind anyone living the active life. It is a characteristic of a man of action, and Tizard was very much a
man of action, that he goes on with his activity until he is
stopped.

He was soon stopped. He was stopped in a somewhat
peculiar fashion. On June 4 he was summoned to see Lindemann at 10 Downing Street. Maddeningly, there is no
record of the conversation; I doubt if anything very direct
was said on either side. The diary simply reports: "June 4.
Thence to see Lindemann at 10 Downing Street. Apparently he had been told by the P. M. to 'drive ahead' with
anything new that may be of use this summer, and there
is enough overlapping of responsibility to hinder almost
anything useful being done."[27]

Tizard must have known that he was out. But the particular way in which he was shown to be out may have
come as a surprise. On June 7 he attended a meeting of
his own Ministry, of which he was still the official scientific adviser—with his own Minister in the Chair. The air
marshals and permanent officials were there. So was Lin-

demann. And it was Lindemann who laid down what the scientific programme should be. Tizard wrote that night: "Doubtful whether S. of S. really expected me. I tried to keep them straight about use of A. I. and G. L. for search-lights—but do not know if I succeeded. I left before the meeting was over as it did not appear that good could be done by staying."[28]

In the next few days Tizard went on with his work and at times saw his friends. A good many of them seem to have thought that a man who had already been proved right so often could not be got rid of so contemptuously.

> Friday, June 21. Meeting at 10 Downing Street to con-sider enemy methods of navigation. P. M. in chair—present Lindemann, S. of S., C. A. S., C. in C's Bomber and Fighter Command, Watson Watt, R. V. Jones and myself. Various decisions reached but would have been reached without those commotions in ordinary way. Afternoon meeting presided over by S. of S. to discuss progress on new developments. As unsatisfactory as previous meeting. Afterwards went to Athenaeum and wrote letter definitely resigning. Showed it to C. A. S. who agreed it was inevitable and asked me to suggest a post of authority for myself. Said this was better left for two or three weeks.[29]

The Chief of Air Staff, Sir Cyril Newell, was, like most of the military people, a devoted supporter of Tizard's. But when they talked of a post of authority, even Tizard, usually clear-sighted, was deluding himself. He was to perform one more first-rate service that year: he was to take part in the classical scientific-military quarrel in 1942;

but, in the sense he had known it, there was to be no more authority for him in that war.

In a few weeks they had thought up something for him to do. Someone, possibly to tempt or mollify him, had revived his old idea of scientific exchange with the United States.

> July 30. A meeting with Fairey in the hall of M. A. P. He said, "I am going to be a member of your staff." I said, "What staff?" He replied that Beaverbrook had just told him that I was to lead a mission to America and that he, Fairey, was to be a member. As Beaverbrook could not see me, Rowlands, the Permanent Secretary of M. A. P., took me to his room and explained that the P. M. wanted me to lead a mission to America for the exchange of technical information . . . I was given a provisional list of "secrets" I could impart, and of information I was to ask for. I said I certainly would not go unless I was given a free hand . . . It looked to me at first sight as rather a neat method of getting a troublesome person out of the way for a time![30]

That was, of course, at least part of the truth. If Tizard had been playing politics he would not have gone. In times of crisis, as all kinds of men have found out, from Trotsky downwards, the first mistake is to absent oneself. But Tizard had always believed in what such a mission could do.

> August 1. Called on Prime Minister at 5.45. Had to wait some time as the Archbishop was with him, which, as the private secretary explained, had quite thrown out

the timetable. The P. M. quite emphatic that the mission was important and that he particularly wanted me to lead it. I asked if he would give me a free hand and would rely on my discretion. He said "of course"—and would I write down exactly what I wanted. So I said I would go, and went into the lobby and wrote out a paper which I left with his secretary. Then I rang up Rowlands and told him that I had accepted and that the P. M. was going to give me full discretion. He said that was quite different from what the P. M. had previously said![31]

Flying the Atlantic in August 1940 meant that a man put his affairs in order. Before he left Tizard arranged that, in case of accidents, his war-time diaries should go to the Royal Society. Those are the diaries from which I have been quoting. He had a proper pride in what he had achieved, and a proper rancour for the way he had been treated. He did not doubt that, if and when competent persons studied the evidence—the diaries and notebooks are full of scientific arguments from 1935 to 1939, which it would not be suitable to quote here—he would get his due.

But no accidents happened, and the mission, on which John Cockcroft was his second in command, was one of the successes of both their lives. American scientists, both at the time and since, have spoken, with extreme generosity, of the effect that visit made. It is true that, mainly because the English had been forced to think in order to survive at all, in most military scientific fields they were ahead. This was preeminently true of radar. Although

English, American, and German scientists had all begun developing radar at about the same time—which incidentally tells one something of the nature of "secret" discoveries—by 1940 the English had carried it further.

Tizard and Cockcroft carried with them a black leather suitcase which Miss Geary, Tizard's secretary, was forced to keep under her bed. She did not know it contained nearly all the important new English war devices—and, of a different order of importance from the rest, the new cavity magnetron. Mr. James Phinney Baxter, writing the story of the American scientific war, has called the black box "the most valuable cargo ever brought to our shores" and "the single most important item in reverse lease-lend." The magnetron, which was invented by Randall and Boot in Oliphant's laboratory at Birmingham, was probably the most valuable single device in the Hitler war.[32] The sight of it set American scientists working all out sixteen months before the United States was in the war at all. As Blackett has said:

> This imaginative act of trust, which Tizard and A. V. Hill first envisaged and finally forced through White-hall, had immensely beneficial effects on the scientific aspects of the allied war effort. Cockcroft reminds us that the mission was magnificently organised by Tizard, and that he had the inspiration to bring a mixed team of serving officers and scientists. For the first time our American friends heard civilian scientists discussing authoritatively the instruments of war, and then heard the Service people following on with practical experience.[33]

When he returned from the mission, Tizard found that he was still out. There was no real job for him. He worked, as a kind of free-lance scientific adviser, in the Ministry of Aircraft Production. Then the R. A. F., which had throughout been loyal to him, put him on the Air Council. But neither of those posts made anything like a full call on his powers. In fact, no post could, while Lindemann was making all the major scientific decisions on the English side of the war.

I saw something of Tizard at the time. He was a very high-spirited man, too high-spirited to be bitter. He was also remarkably free from self-pity. He got a lot of fun out of the solemn paraphernalia of English official life. The dinners at City Companies, the various Boards of Governors of which he was a member—to most of us all that would not have been much consolation, but it was to him. Still, he was only 56, he was at the height of his abilities, he was chafing at the leash. I think he welcomed the final row with Lindemann, not only because he was certain he was right, but also because it gave him something to do.

VIII.

The row occurred in 1942, and it occurred over strategic bombing. We have got to remember that it was very hard for the Western countries to make any significant military effort in Europe that year. The great battles were taking place on the Russian land. So it was natural, and good military sense, that the Western leaders were receptive to any idea for action. It is also true—and this was not such good military sense—that the English and Americans had, for years past, believed in strategic bombing as no other countries had. Countries which had thought deeply about war, like Germany and Russia, had no faith in strategic bombing and had not invested much productive capacity or many elite troops in it. The English had, years before the war began. The strategy had not been thought out. It was just an unrationalised article of faith that strategic bombing was likely to be our most decisive method of making war. I think it is fair to say that Lindemann had always believed in this faith with characteristic intensity.

Early in 1942 he was determined to put it into action. By this time he was Lord Cherwell and a member of the Cabinet, and he produced a cabinet paper on the strategic bombing of Germany. Some cabinet papers are restricted to members of the Cabinet only, and Lindemann occasionally used this technique for circulating a scientific proposal; since he was the only scientist in the Cabinet, discussion was reduced to a minimum. But the paper on bombing went out to the top government scientists.

It described, in quantitative terms, the effect on Germany of a British bombing offensive in the next eighteen months (approximately March 1942—September 1943). The paper laid down a strategic policy. The bombing must be directed essentially against German working-class houses. Middle-class houses have too much space round them, and so are bound to waste bombs; factories and "military objectives" had long since been forgotten, except in official bulletins, since they were much too difficult to find and hit. The paper claimed that—given a total concentration of effort on the production and use of bombing aircraft—it would be possible, in all the larger towns of Germany (that is, those with more than 50,000 inhabitants), to destroy 50 per cent of all houses.

Let me break off for a minute. It is possible, I suppose, that some time in the future people living in a more benevolent age than ours may turn over the official records and notice that men like us, men well-educated by the standards of the day, men fairly kindly by the standards of the day, and often possessed of strong human feelings, made the kind of calculation I have just been describing. Such calculations, on a much larger scale, are going on at this moment in the most advanced societies we know. What will people of the future think of us? Will they say, as Roger Williams said of some of the Massachusetts Indians, that we were wolves with the minds of men? Will they think that we resigned our humanity? They will have the right.

At the time I heard some talk of the famous cabinet paper. I have to say this about my own attitude and that of the people I knew best. We had never had the conventional English faith in strategic bombing, partly on military and partly on human grounds. But now it came to the point it was not Lindemann's ruthlessness that worried us most,[34] it was his calculations.

The paper went to Tizard. He studied the statistics. He came to the conclusion, quite impregnably, that Lindemann's estimate of the number of houses that could possibly be destroyed was five times too high.

The paper went to Blackett. Independently he studied the statistics. He came to the conclusion, also quite impregnably, that Lindemann's estimate was six times too high.

Everyone agreed that, if the amount of possible destruction was as low as that calculated by Tizard and Blackett, the bombing offensive was not worth concentrating on. We should have to find a different strategy, both for production and for the use of élite troops. It fell to Tizard to argue this case, to put forward the view that the bombing strategy would not work.

I do not think that, in secret politics, I have ever seen a minority view so unpopular. Bombing had become a matter of faith. I sometimes used to wonder whether my administrative colleagues, who were clever and detached and normally the least likely group of men to be swept away by any faith, would have acquiesced in this one, as on the whole they did, if they had had even an elementary knowledge of statistics. In private we made the bitter jokes of a losing side. "There are the Fermi-Dirac statistics," we said. "The Einstein-Bose statistics. And the new Cherwell nonquantitative statistics." And we told stories of a man who added up two and two and made four. "He is not to be trusted," the Air Ministry then said. "He has been talking to Tizard and Blackett."

The Air Ministry fell in behind the Lindemann paper. The minority view was not only defeated, but squashed. The atmosphere was more hysterical than is usual in English official life; it had the faint but just perceptible smell of a witch hunt. Tizard was actually called a defeatist. Strategic bombing, according to the Lindemann policy, was put into action with every effort the country could make.

The ultimate result is well known. Tizard had calculated that Lindemann's estimate was five times too high. Blackett had put it at six times too high. The bombing survey after the war revealed that it had been ten times too high.

After the war Tizard only once said "I told you so." He gave just one lecture on the theory and practice of aerial bombing. "No one thinks now that it would have been possible to defeat Germany by bombing alone. The actual effort in manpower and resources that was expended on bombing Germany was greater than the value in manpower of the damage caused."

During the war, however, after he had lost that second conflict with Lindemann, he went through a painful time. It was not easy, for a man as tough and brave as men are made, and a good deal prouder than most of us, to be called a defeatist. It was even less easy to be shut out of scientific deliberations, or to be invited to them on condition that he did not volunteer an opinion unless asked. It is astonishing in retrospect that he should have been offered such humiliations. I do not think that there has been a comparable example in England this century.

However, the Establishment in England has a knack of looking after its own. At the end of 1942 he was elected to the presidency of Magdalen College, Oxford. This is a

very honourable position, which most official Englishmen would accept with gratitude. So did Tizard. There
are no continuous diary entries at this period, although
now he had plenty of time. For once his vitality seems to
have flagged.

I think there is little doubt that, sitting in the Lodgings
at Magdalen during the last thirty months of war, he often thought of Whitehall with feelings both of outrage
and regret. Here he was, in one of the most splendid of
honorific jobs, but his powers were rusting—powers that
were uniquely fitted for this war. He knew, more accu-
rately than most men, what he was capable of. He believed, both in his dignified exile in Oxford and to the end
of his life, that if he had been granted a fair share of the
scientific direction between 1940 and 1943, the war might
have ended a bit earlier and with less cost. As one goes
over the evidence it is hard not to agree with him.

After the war, he and Lindemann were never reconciled.
In Whitehall they performed a Box and Cox act which
had a note of sarcastic comedy. In 1945, with the political
defeat of Churchill, Lindemann went back to his professorial chair at Oxford. Tizard was promptly invited by the
Labour Government to become chairman of the Advisory
Council on Scientific Policy, and also of the Defence Research Policy Committee, that is, to become the government's chief scientific adviser, very much in the mode
that Killian and Kistiakowsky have been employed in the
United States. In 1951 Churchill and Lindemann returned
to power. Tizard rapidly resigned.

It caused a good deal of comment that Tizard was

never put in the House of Lords, but that did not trouble him. The only thing he was known to grumble about was his pension, which, as I previously mentioned, was derisory. In his very last years, when he and Lindemann were both getting old, he had to take some directorships to make money for himself and provide for his wife. Lindemann died in 1957. Tizard outlived him by two years.

IX.

There ends my cautionary story. Now I want to suggest just which cautions we can reasonably extract from it. First we have got to allow for those features of English government and administration which are peculiar to us. There are some features which do not travel, which are inexplicable and boring to Americans and Russians involved in their own problems of science and government. These features are, as American publishers used to say in pained tones of English novels, too British. The chief of them, I think, is the small size, the tightness, the extreme homogeneity, of the English official world. I. I. Rabi once told me that, on his first visit to England in wartime, I believe in 1942, he found Churchill actually handling the prototype of a new radar set in No. 10 Downing Street. Rabi wondered, why did the English insist on running the war as though it were a very small family business?

It is perfectly true that the English unconsciously adopt all sorts of devices for making their population, genuinely small by world standards, seem a good deal smaller than it really is: just as the United States, it seems to me, does exactly the reverse.

But, though that is true, I do not think it affects the major lessons of my story. There is a great deal in closed politics which is essentially the same in any country and in any system. If we are going to begin to understand what goes on, and so do better, I am sure it is wise to take for granted that other countries are much the same as ourselves, not vastly different. To a friendly observer, it often seems that Americans endanger themselves most when they get most possessed by a sense of their own uniqueness. In all the problems I am now discussing, government science, closed politics, secret choices, there is no such uniqueness.

In these matters, by the sheer nature of the operations, all countries have to follow very similar laws. No country's governmental science is any "freer" than any other's, nor are its secret scientific choices. I beg you to listen to this. It is said by someone who knows you a bit, who loves you a lot, and who is passionately anxious to see your generous creative forces set loose in the world. You have no special advantages in this domain of science and decision. Listening to American and Soviet scientists, trying to study the way in which you both do your government science, I am struck, not by the differences, but by the similarities. If there is any difference, it is perhaps that, because of the special privileges and autonomy of the Soviet Academy, Russian scientists take a slightly loftier attitude: and also, though this may be a superficial impression, I fancy their major choices involve more scientific minds, are slightly more broadly based, than with you or us.

So I believe we are in the same boat and that all countries can learn from each other's concrete experience. We all know the ideal solutions. First, you can abolish some, though probably not all, secret choices as soon as you abolish nation-states. Second, the special aura of difficulty and mystery about these choices will at least be minimised as soon as all politicians and administrators are scientifically educated, or at any rate not scientifically illiterate. Neither of these ideal solutions is in sight. We may therefore not be entirely wasting our time if we try to analyse some phenomena of scientific choice in "closed" politics.

I have used the phrase "closed politics" before. I mean any kind of politics in which there is no appeal to a larger assembly—larger assembly in the sense of a group of opinion, or an electorate, or on an even bigger scale what we call loosely "social forces." For instance, some of the struggles in an English Cabinet partake of the nature of closed politics: but this is not pure closed politics, since the Prime Minister or any member can if pressed move from personal to mass opinion. On the other hand, almost all the secret scientific choices are something like pure closed politics.

In my type-specimen, during the whole of his conflicts with Lindemann, Tizard had no larger body of support to call on. If he had been able to submit the bombing controversy[35] to the Fellows of the Royal Society, or the general population of professional scientists, Lindemann would not have lasted a week. But of course Tizard could do no such thing: and that is true of most conflicts in government science and of all secret choices.

So we find ourselves looking at the classical situations of closed politics. The most obvious fact which hits you in the eye is that personalities and personal relations carry a weight of responsibility which is out of proportion greater than any they carry in open politics. Despite appearances, we are much nearer than in ordinary government to personal power and personal choice. A crude result is that, at this moment, all countries are not unlikely to be at the mercy of scientific salesmen.

In the Tizard-Lindemann story, we saw three of the characteristic forms of closed politics. These three forms are not often completely separable, and usually fuse into each other, but they are perhaps worth defining. The first is committee politics. There is, of course, a complex morphology of committee politics, and everyone who has ever lived in any society, in a tennis club, a factory dramatic group, a college faculty, has witnessed some of its expressions. The archetype of all these is that kind of committee where each member speaks with his individual voice, depends upon his personality alone for his influence, and in the long run votes with an equal vote.

The Tizard Committee itself was a good example. The members did not represent anyone but themselves. Their only way of affecting conclusions was by their own mana and their own arguments. If it came to a disagreement, then the ultimate decision, which any official committee leans over backwards to avoid, was by means of "counting heads." That was what happened, though the circumstances were dramatic, when Lindemann was opposing Tizard over the priority for radar. Everyone round the table knew that it was three to one against Lindemann.[36] In this archetype of a committee, with personalities of approximately equal toughness, with no external recourse except a Churchill out of power and so possessing only nuisance value, that meant his case was lost.

I have just said that any official committee, certainly any English official committee, is reluctant about taking an open vote. I believe that such a vote has never in fact been taken in the English Cabinet: but of course the substance of a vote, the way opinion has divided, is obvious enough. If you want open votes, so as to see the committee operation in its full beauty, you need to go to societies which do not damp down the friction of personalities— such as the smaller colleges of my own Cambridge, which cheerfully proceed to open votes on all sorts of controversies, including personal appointments. I suppose the most famous open vote of this century happened when, in October 1917, smuggled for safety into the house of a political enemy, Lenin moved his resolution to the Central Committee of the Bolshevik party "That . . . [very long parenthesis defining the conditions] . . . the Bolsheviks do now seize power." The voting was ten to two in favour, with Kameniev and Zinoviev voting against.

There is nothing, by the way, in committee politics which is specially connected with American or English parliamentary institutions. The Venetian oligarchy were great masters of committee work and carried out most of their government by its means. The Council of Ten (which usually sat as a body of seventeen) and the Heads of the Ten (who were an inner committee of three) made most of the executive decisions. I doubt if there is much that any of us could have taught them about committee politics. In a book of mine some years ago I wrote about a meeting of high officials:

> These men were fairer, and most of them a great deal abler, than the average: but you heard the same ripples below the words, as when any group of men chose anyone for any job. Put your ear to those meetings and you heard the intricate, labyrinthine and unassuageable rapacity, even in the best of men, of the love of power. If you have heard it once—say, in electing the chairman of a tiny dramatic society, it does not matter where—you have heard it in colleges, in bishoprics, in ministries, in cabinets: men do not alter because the issues they decide are bigger scale.[37]

I should still stand by each word of that.

The second form of closed politics I think I had better call "hierarchical politics"—the politics of a chain of command, of the services, of a bureaucracy, of a large industry. On the surface these politics seem very simple. Just get hold of the man at the top, and the order will go down the line. So long as you have collected the boss, you have got nothing else to worry about. That is what people believe—particularly people who are both cynical and unworldly, which is one of my least favourite combinations—who are not used to hierarchies. Nothing could be more naive.

Chain-of-command organisations do not work a bit like that. English organisations, our Civil Service, our armed Services, are moderately well disciplined, by existing standards. Certainly our serving officers do not show the same enthusiasm for publicising their point of view, especially when they cut across higher authority, as some American officers appear to show. But, in reality, though not on the surface, both our countries work much the same way.

To get anything done in any highly articulated organisation, you have got to carry people at all sorts of levels. It is their decisions, their acquiescence or enthusiasm (above all, the absence of their passive resistance), which are going to decide whether a strategy goes through in time. Everyone competent to judge agrees that this was how Tizard guided and shoved the radar strategy. He had the political and administrative bosses behind him from the start (Churchill and Lindemann being then ineffective). He had also the Air Staff and the Chiefs of Command. But he spent much effort on persuading and exhorting the junior officers who would have to control the radar chains when they were ready.

In the same way, he was persuading and exhorting the scientists who were designing the hardware, and the administrators who had to get it made. Like all men who understand institutions, Tizard was always asking himself the questions "Where to go to? For which job?" Often, for a real decision as opposed to a legalistic one, the chap who is going to matter is a long way down the line. Administrators like Hankey and Bridges were masters of this kind of institutional understanding, and they were able to prod and stroke, caress and jab, the relevant parts of the English organism, so that somehow or other, in a way that made organisational diagrams look very primitive, the radar chain got made.

I remember myself, very early in the war, being sent for by a high functionary, much to the bafflement and, I am afraid, to the irritation, of my official superiors. I was a junior official, having gone in as a temporary a few months before: but I had taken on myself the job of producing large numbers of radar scientists. As usual, everyone had forgotten the sheer human needs, in terms of numbers of trained minds, of a new device. I got my summons and went off to the Treasury. My interlocutor was so many steps above me in the hierarchy that no regular communication was possible. That did not matter. Later on, we became friends. The interview, however, took about five minutes. Was this scheme going all right? Should we get enough men? At the right time? The answer to those questions was yes. Did I need any help? No, not just then. That was all. That is the way hierarchical politics sometimes has to work. Granted a serious objective, granted a long-term and unspoken respect for certain rules, it often works very well.

This is a form of politics which has not yet received the attention it needs, if one is going to have any feel, not for how an elaborate organisation is supposed to operate, but for how it does in fact.[38] It cuts across all kinds of romantic stereotypes of official power. The top bosses of great corporations like General Motors, or General Electric, or their English equivalents, could not act even if they wanted to, could not act by the intrinsic nature of their organisation, like the proprietors of a small film company. Blissful expressions of power, such as hire and fire, get more remote from reality the more elaborate your organisation is, and the nearer you are to the top of it. I suspect that hierarchical politics are probably more interesting and complex in the United States than in any country in the world, certainly more interesting than in any country in the West.

The third form of politics in the Tizard-Lindemann story is the simplest. I shall call it "court politics." By court politics I mean attempts to exert power through a man who possesses a concentration of power. The Lindemann-Churchill relation is the purest example possible of court politics.

In 1940, as I described it, Lindemann asked Tizard to call on him at 10 Downing Street. At that time Tizard was the most senior scientific adviser in government employment. Lindemann had no official position whatever; he was the confidential friend of Churchill. Before the end of their conversation Tizard knew that his authority was over. Within three weeks he had resigned.

For another eighteen months, until the end of 1941, Lindemann still held no official position whatever: but he had more direct power than any scientist in history. Roosevelt had a court too, and there must have been a lot of court politics throughout his administrations; but, so far as I know, no scientist ever got near to being intimate with him, and Vannevar Bush and his colleagues were operating at the ordinary official distances and through the ordinary official techniques. Hitler had a court, but he, to an extent quite unparalleled, kept the power to himself. Incidentally, no scientist seems to have got anywhere near him, though he was interested in weapons. His total lack of scientific comprehension was fortunate for the world.

Churchill and Lindemann, however, really did work together, on all scientific decisions and on a good many others, as one mind. In his early days as grey eminence to the Prime Minister, Lindemann made it obvious, by holding his interviews in 10 Downing Street or by threatening Churchill's intervention. Very soon this was not necessary. Bold men protested to Churchill about Lindemann's influence,[39] and were shown out of the room. Before long everyone in official England knew that the friendship was unbreakable, and that Lindemann held real power. Before long also men had accustomed themselves to that degree of power and jumped up behind it; for an overwhelming majority of men find a fascination in seeing power confidently used, and are hypnotised by it. Not entirely through self-seeking, though that enters too.

The fact that the bombing policy was forced through with so little opposition is a typical example of the hypnosis of power. A good many men read the Tizard and Blackett papers. A certain proportion felt, men being men, that, if a scientific statesman like Tizard could be ignominiously swept aside, lesser persons had better keep quiet. It is very easy, in an atmosphere of crisis, in the midst of secret decisions, for men to surrender both their reason and their will. I can still hear someone, a man normally tough and intelligent, saying to me one black night: "The P. M. and Prof. have decided—and who are we to say them nay?"

Judged by the simple criterion of getting what he wanted, Lindemann was the most successful court politician of the age. One has to go back a long way, at least as far as Père Joseph, to find a grey eminence half as effective. Incidentally, there exists a romantic stereotype of the courtier—as someone supple, devoid of principle, thinking of nothing except keeping his place at court. Now Lindemann was, in functional terms, a supreme courtier; and yet no one could be more unlike that stereotype. Life is not as simple as that, nor as corrupt in quite that way. Throughout his partnership with Churchill, Lindemann remained his own man. A remarkable number of the ideas came from him. It was a two-sided friendship. There was admiration on Lindemann's side, of course, but so there was on Churchill's. It was a friendship of singular quality—certainly the most selfless and admirable thing in Lindemann's life, and in Churchill's, much richer in personal relations, it nevertheless ranked high. It is ironical that such a friendship, which had much nobility and in private showed both men at their human best, should in public have led them into bad judgments.

In all closed politics the three forms I have isolated—committee politics, hierarchical politics, court politics—

interweave, interact, and shift from one to the other.[40] That is independent of the objectives, which may be good or bad; it is simply the way men have to operate, in order to get anything done at all. I do not mean that as satire. *Satire is cheek.*[41] It is the revenge of those who cannot really comprehend the world or cope with it. No, I mean my description of politics to be taken as neutral statements. So far as I have been able to observe anything, this is how the world ticks—not only our world, but also the future world one can imagine, juster and more sensible than ours. It seems to me important that men of good will should make an effort to understand how the world ticks; it is the only way to make it tick better.

X.

After looking at the Tizard-Lindemann story, and reflecting a bit on the kinds of politics, can we find any guide to action? Is there any way, in this great underground domain of science and government, in which we can arrange to make choices a little more reasonably?

Let me say at once that I have no easy answers at all. If there were any easy answers, they would have been found by now. The whole problem is an intractable one, one of the most intractable that organised society has thrown up. It is partly the expression, in political and administrative terms of the split between two cultures that I have said something about elsewhere.[42]

But, though the answers have not presented themselves, I think we have advanced far enough to know certain things to avoid. We know some of the sources of bad judgments and bad choices. I think most of us would agree that it is dangerous to have a solitary scientific overlord. It is specially dangerous to have him sitting in power, with no scientist near him, surrounded by politicians who think of him, as some of Churchill's colleagues thought of Lindemann, as the all-wise, all-knowing Prof. We have seen too much of that, and we should not like it to happen again.

And yet, as I say that, I wonder if I am becoming too cautious, too much in love with an old country's predilection for checks and balances. Lindemann made some bad choices, but he also drove some things through as a nonscientist could not have done. Imagine that, in that same position of solitary scientific power, Tizard had been installed: or that Vannevar Bush had been as close to Roosevelt as Lindemann was to Churchill. In either of those cases the positive good would have been startling. Still, I do not think it is overcautious to remember that that has never happened. The chances of getting a Tizard or a Bush as scientific overlord are pretty remote. On the whole, I am still inclined to believe that the obvious dangers outweigh the vestigial possibility of good.

That is fairly clear. We ought not to give any single scientist the powers of choice that Lindemann had. It is even clearer, in my mind at least, that there is a kind of scientist to whom we ought not to give any power of choice at all. We have seen some examples of how judgments were distorted, enough to specify some of the people to fight shy of. Various kinds of fear distort scientific judgments, just as they do other judgments: but, most of all, the self-deceiving factor seems to be a set of euphorias. The euphoria of gadgets; the euphoria of secrecy. They are usually, but not invariably, combined. They are the origin of 90 per cent of ill-judged scientific choices. Any scientist who is prone to these euphorias ought to be kept out of government decisions or choice-making, at almost any cost. It doesn't matter how good he is at his stuff. It doesn't matter if the gadgets[43] are efficacious, like the atomic bomb, or silly, like Lindemann's parachute mines for dropping on airscrews.[44] It doesn't matter how confident he is; in fact, if he is confident because of the euphoria of gadgets, he is doubly dangerous.

The point is, anyone who is drunk with gadgets is a menace. Any choice he makes—particularly if it involves comparison with other countries—is much more likely to be wrong than right. The higher he climbs, the more he is going to mislead his own country.

The nearer he is to the physical presence of his own gadget, the worse his judgment is going to be. It is easy enough to understand. The gadget is *there*. It is one's own. One knows, no one can possibly know as well, all the bright ideas it contains, all the snags overcome. I have felt something like it at second hand, over gadgets I have seen developed. Seeing the first English jet flying in 1942, I could not believe this was not unique. It was like denying one's own identity to credit there was anything else like that in existence. As a matter of fact, of course, there were in existence quite a lot like that. The Germans had already got a jet flying even more impressively. In cold blood the probabilities dawn again, just as they dawned upon anyone connected with radar, who found the same gadgets being developed in the same loving secrecy in England, in the United States, in Germany and elsewhere.

The overriding truth is a bleak one, if one is living in the physical presence of gadgets and spends one's creative force developing them: that societies at about the same level of technology will produce similar inventions. In military technology in particular, where the level of the United States and the U.S.S.R. is very much the same and where the investment of scientists and money is also similar, it would be astonishing if either society kept for long anything like a serious, much less a decisive, technical lead.

It is overwhelming odds that one country will get its nose in front in one field for a short time, the other somewhere else. This situation, fluctuating in detail but steady in the gross, is likely to continue without limit. It is quite unrealistic, and very dangerous, to imagine that the West as a whole can expect a permanent and decisive lead in military technology over the East as a whole. That expectation is a typical piece of gadgeteers' thinking. It has done the West more harm than any other kind of thinking. History and science do not work that way.

If one is not existing in the immediate presence of gadgets, it is a little less impossible to keep a kind of rudimentary common sense. The news of the first atomic pile reached a few of us in England in 1943. In the somewhat inelegant language of the day, we knew the atomic bomb was on. We heard people, intoxicated by the discovery, predicting that it would give the United States unheard-of power for so long as one could foresee. We did not believe it. We had no special prescience, but we were outside the area of euphoria. We speculated on how long it would take a country with the scientific and technical capacity of Russia to catch up, once the discovery was known. We guessed about six years. We were wrong. One always over-estimates these periods. It took them four.

It is one of the firmest convictions of most of the best administrators I have known that scientists, by and large, could not do their job. There are many reasons for this conviction, including various human frailties, and I shall return to it at the end. But there is one good one. Many administrators have had to listen to the advice of scientist-gadgeteers. To Bridges and his colleagues, to a good many of the high civil servants who played a part in the Tizard-Lindemann story, it must have appeared scarcely human that men should be so lacking in broad and detached judgment.[45] Most administrators would go on to feel that there is something of the gadgeteer hiding in every scientist.

I have to admit that there is something in it. I should phrase it rather differently. The gadgeteer's temperament is an extreme example of a common scientific tempera-ment. A great many scientists have a trace of the obses-sional. Many kinds of creative science, perhaps most, one could not do without it. To be any good, in his youth at least, a scientist has to think of one thing, deeply and obsessively, for a long time. An administrator has to think of a great many things, widely, in their interconnections, for a short time. There is a sharp difference in the intellectual and moral temperaments. I believe, and I shall lay some stress on this later, that persons of scientific education can make excellent administrators and provide an element without which we shall be groping: but I agree that scientists in their creative periods do not easily get interested in administrative problems and are not likely to be much good at them.

The euphoria of secrecy goes to the head very much like the euphoria of gadgets. I have known men, prudent in other respects, who became drunk with it. It induces an unbalancing sense of power. It is not of consequence whether one is hugging to oneself a secret about one's own side or about the other. It is not uncommon to run across men, superficially commonplace and unextravagant, who are letting their judgment run wild because they are hoarding a secret about the other side—quite forgetting that someone on the other side, almost indistinguishable from themselves, is hoarding a precisely similar secret about them. It takes a very strong head to keep secrets for years, and not go slightly mad. It isn't wise to be advised by anyone slightly mad.

XI.

I could go on accumulating negatives and empirical prescriptions. We know something about what not to do and whom not to pick. We can collect quite a few working tips from the Tizard-Lindemann story. For instance, the prime importance, in any crisis of action, of being positive what you want to do and of being able to explain it. It is not so relevant whether you are right or wrong. That is a second-order effect. But it is cardinal that you should be positive. In the radar struggle Tizard and his committee were positive that theirs was the only hope, and Lindemann had only quibbles and fragmentary ideas to set against it. Over bombing, Lindemann was positive that he had the recipe to win the war. Tizard was sure he was wrong, but had nothing so simple and unified to put in its place. Even at the highest level of decision, men do not really relish the complexity of brute reality, and they will hare after a simple concept whenever one shows its head.

We also saw that a committee like the Tizard Committee is, in the right conditions, as sharp a tool for doing business as government can find. What are the conditions? As a sighting shot I should say:

(1) The objective must be clear and not too grandiloquently vast. A scientific committee set to advise on the welfare of all mankind is not likely to get very far. The objective of the Tizard Committee—to defend England in a foreseeable short-term future against air attack—is about as much as anyone can hope actually to cope with.

(2) The committee has got to be "placed" within the government structure. It is usually not difficult to do this, if one has people who know the government machine (or organism, since machine is a bad word) by touch. Different government machines need a different touch, and as a rule a foreigner, however well he knew the

country, would dither about where the optimum place should be. To fit the local English structure, the Tizard Committee could not have been better placed, partly by good management, partly by good luck. It was not so high as to get out of touch with the working administrators and the serving officers, or to arouse too much envy (very important in a compact country). But it had its own links with ministers and top civil servants. In the United States, if I have not got it wrong, there is not the same problem of fitting into a highly organised and very powerful civil service. On the other hand, the committee has to survive in a welter of constitutional and contractual complications, much more elaborate than any the English know. As for the Soviet Union, I have an impression that the correct placing would bring in a good many questions of academic status.

(3) To be any real good, the committee has to possess (or take, as the Tizard Committee took) powers of action. It needs, at the least, the power of inspection and follow-up. If it does not have those, it will be too far from the reality it is trying to decide about, and too far from the people who are supposed to carry out the decisions. Advisory committees, if they are confined to pure advice and never get near the point of action, fade away into a kind of accidie.

As a matter of historical fact, these conditions for an effective committee have quite often been achieved. In any particular case, it ought to be reasonably easy to achieve them again. It is—and this is bad luck for us all—specially easy to do this for military objectives. Military objectives are nearly always more precise than benevolent ones: which is why military technology has been easier for ingenious men to think about.

Again unfortunately, the constraints of secrecy, though they disturb the comparative judgment, do not disturb the scientific process. In more liberal days, in the days of Rutherford's Cambridge, Bohr's Copenhagen, Franck's Gottingen, scientists tended to assume, as an optimistic act of faith, as something which ought to be true because it made life sweeter, that science could only flourish in the free air.

I wish it were so. I think everyone who has ever witnessed secret science and secret choices wishes it were so. But nearly all the evidence is dead against it. Science needs discussion, yes: it needs the criticism of other scientists: but that can be made to exist, and of course has been made to exist, in the most secret projects. Scientists have worked, apparently happily, and certainly effectively, in conditions which would have been thought the negation of science by the great free-minded practitioners. But the secret, the closed, the climate which to earlier scientists would have been morally intolerable, soon becomes easy to tolerate. I even doubt whether, if one could compare the rate of advance in one of the secret sciences[46] with one of those which is still open to the world, there would be any significant difference. It is a pity.

There is a difference, though, in the rate at which the sciences open to the world get into action. Since those sciences are by definition the ones which cannot be pointed at a military objective, they get into action slower.

The exceptions, though perhaps only partial exceptions, are the cluster of sciences which can be applied to medicine. In medicine the objectives are often as clear-cut as in military science.[47] In fact, there is a certain grim family resemblance. This gives edge and sharpness to the deployment of medical research. For it is not the nature of the objective that makes for speedy action, whether it is destructive or on the side of life. All that matters is that there should be an objective at all.

I am speaking very much as an outsider here, and even if I were not, it is difficult to be sure what one means when one speaks of the efficiency of research and development. But, if that phrase means anything, I should have thought the efficiency of medical research in both the United States and England is a good deal higher than of military research. The choices, often because they are not so much all-or-nothing, have been more sensibly made. This is true, although the administrative techniques in the two countries are not the same. Our Medical Research Council, working with funds Americans would think derisory, is an unusual example, very much admired among people who are studying the arts of government, of a government organ which is acting not so much as a controlling force, but as an impresario.

So in military science, and on a lesser scale in medical science, government manages to get some results. But an awful lot of life doesn't consist either of trying to accelerate people's deaths or alternatively to delay them. In the application of science to this vast mid-range of human life, the problems are vaguer, the impetus is less, the pressures of government do not weigh so heavy. A good many benevolent initiatives get lost, although government in the United States, and with slightly less conviction in England, might think that (*a*) this was not their business, (*b*) the initiatives will work their way out elsewhere in the society. It is arguable that that is so, but I am by no means convinced. And governments are not convinced either, because they have set out some sort of springboard where these initiatives can get started. In the United States, unless I am wrong, this springboard ought to be provided by the National Research Council. In England, by the Advisory Council on Scientific Policy. In the Soviet Union, by the Academy of Sciences itself, which is a much smaller body than the U.S. National Academy of Sciences or than the Royal Society of London. The Soviet Academy of Sciences is made up of something like 250 full Academicians, and about 150 corresponding members. It contains historians, economists, various kinds of literary scholars, and even creative writers. About 70 per cent are scientists in the restricted Western sense. It is difficult to guess how completely they succeed as a source of scientific initiative. As for us, I do not think anyone would claim that our organs are well-designed for the job.

Does that matter? Is there a job? Hasn't the West in particular got so much applied science in so many quarters that it doesn't need any encouragement?

Does anyone in his senses need more material possessions than the ordinary comfortably off professional American? Or indeed as many? I have some sympathy with anyone who asks me that. And yet, with the ultimate attitude behind it, I haven't so much sympathy after all.

Why not leave well alone? You have said yourself that not many scientists make good administrators. Why worry about science and government? Why not keep the scientists in their place, as we used to, and just call them out to give advice to wiser men?

Isn't the first, the only serious problem of our time, to save the peace? Why does it matter what we do with the scientists? Isn't it the statesman's job to save the peace? What does it matter about scientists?

I am familiar with those questions. They are asked by intelligent men. There is a lot of truth in some of them. And yet they are no good. Or rather, they spring from roots from which spring also many of our dangers and our losses of hope. One of those dangers is that we are beginning to shrug off our sense of the future.

This is true all over the West. True even in the United States, though to a lesser extent than in the old societies of Western Europe. We are becoming existential societies—and we are living in the same world with future-directed societies. This existential flavour is obvious in our art. In fact, we are becoming unable to accept any other kind of art. It is there to be seen in quarters much nearer the working mechanism of our society, in the deepest of our administrative arrangements, in the way we make the secret choices that I spoke of at the beginning, in the nature of the secret choices themselves. We seem to be flexible, but we haven't any model of the future before us. In the significant sense, we can't change. And to change is what we have to do.

That is why I want scientists active in all the levels of government. By "scientists" here I mean people trained in the natural sciences, not only engineers, though I want them too. I make a special requirement for the scientists proper, because, partly by training, partly by self-selection, they include a number of speculative and socially imaginative minds. While engineers—more uniform in attitude than one would expect a professional class to be—tend to be technically bold and advanced but at the same time to accept totally any society into which they may have happened to be born. The scientists proper are nothing like so homogeneous in attitude, and some of them will provide a quality which it seems to me we need above everything else.

I do not merely mean here that, if we had scientists of any kind diffused through government, the number of people helping to influence secret choices is bound to increase. That is true. In my view, and it is one of the points from which I started, it would be a real gain. It is a clear advantage to the Soviet Union that they have, right at the top of the political and administrative trees, a fairly high proportion of men with scientific or technical training. The proportion of these men in the top executive organs, or among high-ranking diplomats, seems to be somewhere between 35 and 45 per cent, which is far higher than in the United States or England. In the fields where they have made better technical choices than either of us, and there are plenty, this collective influence has no doubt been a help. But, though that is a real gain, it is secondary to what I have most in mind. I believe scientists have something to give which our kind of existential society is desperately short of: so short of, that it fails to recognise of what it is starved. That is foresight.

I am not saying, of course, that all scientists have foresight and no one else has. Foresight is a fairly rare quality. Mr. Secretary Stimson showed some of it, more than other political figures at the time, in his memorandum to President Truman, dated April 25, 1945, about the consequences of the atomic bomb.[48] But compare the kind of prescience in this memorandum with that of Franck and the Chicago scientists in their famous letter ten weeks later.

Stimson had to rely on his political sense. Franck and his colleagues had training and something which we can loosely call knowledge behind them. It was not quite knowledge. It was much more an expectation of knowledge to come. It was something that a scientist, if he has this kind of sensitivity latent in him, picks up during his scientific experience.

I believe it is something we grossly undervalue: rather like paleolithic men, before arithmetic had been invented, jeering at someone who had a knack of counting on his fingers. I suppose most scientists possess nothing of this foresight. But, if they have any trace of the capability, then their experience, more than any experience at present open to us, gives them the chance to bring it out. For science, by its very nature, exists in history. Any scientist realises that his subject is moving in time—that he knows incomparably more today than better, cleverer, and deeper men did twenty years ago. He knows that his pupils, in twenty years, will know incomparably more than he does. Scientists have it within them to know what a future-directed society feels like, for science itself, in its human aspect, is just that.

That is my deepest reason for wanting scientists in government. I have tried a shot at an explanation why in their youth they are often not good at the arts of administration. As one thinks back to the operations of the Tizard Committee, it is worth remembering that their decisions were carried out by professional administrators. If these had been replaced by scientists, the scientists would almost certainly have done worse.

But that is only half of it. I spent twenty years of my life in close contact with the English professional administrators. I have the greatest respect for them—more respect, I think, than for any professional group I know. They are extremely intelligent, honourable, tough, tolerant, and generous. Within the human limits, they are free from some of the less pleasing group characteristics. But they have a deficiency.

Remember, administrators are by temperament active men. Their tendency, which is strengthened by the nature of their job, is to live in the short term, to become masters of the short-term solution. Often, as I have seen them conducting their business with an absence of fuss, a concealed force, a refreshing dash of intellectual sophistication, a phrase from one of the old Icelandic sagas kept nagging at my mind. It was: "Snorri was the wisest man in Iceland who had not the gift of foresight."[49]

Foresight in this quotation meant something supernatural, but nevertheless the phrase stayed with me. The wisest man who had not the gift of foresight. The more I have seen of Western societies, the more it nags at me. It nags at me in the United States, just as in Western Europe. We are immensely competent; we know our own pattern of operations like the palm of our hands. It is not enough. That is why I want some scientists mixed up in our affairs. It would be bitter if, when this storm of history is over, the best epitaph that anyone could write of us was only that: "The wisest men who had not the gift of foresight."

APPENDIX

"Whoever he is, whether he is the wisest scientist in the world, we must never tolerate a scientific overlord again."

The Godkin Lectures on *Science and Government* were delivered in December 1960 and published in April 1961. They make quite a short book, not more than 20,000 words or so. There have been a great many more words, however, in the shape of discussions, reviews, articles, and private communications—probably more than in the original piece. Then in October 1961 the four-volume history of the strategic bombing offensive, *The Strategic Air Offensive against Germany* 1939–1945 (H. M. Stationery Office), by the late Sir Charles Webster and Dr. Noble Frankland, appeared at last. These volumes are an essential source for anyone interested in the themes I myself was writing about. They are, of course, much more than that. In their turn they received a remarkable amount of comment and controversy, far more, which was natural and right, than my little book had done. Finally, in November 1961, Lord Birkenhead published his official biography of Lord Cherwell (F. A. Lindemann) called *The Professor and the Prime Minister.* More arguments started: sometimes the same arguments repeated, sometimes new ones, now the amount of material open to inspection had become greater.

We can now study a good many more documents than I was able to when I first prepared my lectures. It may be some years before we are given any new documents of importance. There are still two official caches which, when broken open, will clear up some of the areas of doubt. I will mention these potential sources later, but it is unlikely that they will do more than sharpen the knowledge we already possess.

Other official sources are at present being explored (for example, the history of the British side of atomic energy). Some of these may give more spectacular results. So may domains of the scientific war, particularly, those connected with the army, which so far as I know are still being left in obscurity. It is always possible that we shall have memoirs or autobiographies from some of the major scientists of the war, as, if he had lived longer, we should have had Tizard's.

For the present, however, we must make do with what we have. It ought to be enough to bring us, on most issues, to sensible conclusions. Some of the arguments will remain, and some of the mysteries. But these will not look so numerous, once the dust has settled. With every new batch of documents so far, acrimonious words have been exchanged. I don't intend to add to these. I shall permit myself one subdued grumble on my own account: but that apart—and it does not matter much anyway— I have no doubt what is the most useful thing I can do. I have asked myself a simple question. There is now a lot of information which was not published when I gave the Godkin Lectures. If I had had this information, how would it have affected what I wrote?

II.

I must not raise false expectations. If I were doing it again, I should say, in essence, the same things. I should alter it a little in emphasis, a little in tone, I should make rather more qualifications, but in the end I should come, with an even greater sense of urgency, to the same conclusions. The main lessons I wanted to draw seem to me, after almost twelve months' preoccupation with scientific-military history, above all with the history of the strategic bombing offensive, to matter more vitally than they ever did; and I am using the word "vitally" in its original and naked sense. I regret only that I did not make those lessons clearer. By a failure of literary tact, I allowed the anecdote or parable of Tizard-Lindemann to distract too much attention from the lesson it was supposed to teach. If I were doing it again, I should try to get the lesson and the parable into better proportion. But I should still use the parable. It is the best example in our time of what I was trying to explain, and of what we ought to know. In the light of what I have learned, I should introduce some amendments and speculations: but, again in the light of what I have learned, the anecdote seems to me not weaker, but stronger—sadder, harsher, but more illuminating.

The original lectures can, for convenience, be divided into five parts: (a) the statement of the problem, that is, the problem of decision-making in "closed" politics, in particular, decision-making in scientific affairs; (b) the character, scientific qualities, and achievements of Tizard;

(c) the character, scientific qualities, and achievements of Lindemann; (d) the Tizard-Lindemann conflicts; (e) the lessons, negative and positive, and the problem of scientific judgment, in particular scientific judgment as applied to military operations. I will now try to have another look at each of these in turn.

III.
Decision-Making in Closed Politics

I wish that on pages 1 and 2 I had said much more and said it more sharply. The longer I think about the way decisions have been taken, are being taken, and will continue to be taken, the more frightened I get. I have been accused, by friendly critics, of liking closed politics too much. In one sense there is something in what they say. I have devoted a lot of attention and interest to closed politics: I understand them pretty well: and at times in my life I have been at home there. It is a fairly short step from being at home with something to liking it more than one should. There is a genuine corruption of familiarity.

All that is true. Yet, the more I learn about these methods of taking decisions, the more the dangers appall me. Not only the practical dangers, but the moral dangers too. I do not believe that any person, moderately detached and moderately sentient, can read the sombre pages of *The Strategic Air Offensive* without a feeling of something like vertigo—that *this* is the way we have to make cardinal choices. The Chief of Air Staff on February 15, 1942, writes a minute:

> Ref the new bombing directive: I suppose it is clear that the aiming points are to be the built-up areas, *not*, for instance, the dockyards or aircraft factories where these are mentioned in Appendix A.
>
> This must be made quite clear if it is not already understood.[1]

As a piece of information, what percentage of Americans and English realised that this was their countries' intention? It is a very interesting example of collective moral responsibility. In the long future, perhaps the history of our times, and our methods of making war, will be written by some Asian Gibbon: if so, *The Strategic Air Offensive* will provide him with a good many of his most sardonic laughs.

How else can we make this type of decision, though, so long as we have nation-states? It would be nice to think that the United States and the Soviet Union might suddenly renounce all scientific-military decisions taken in secret; it would be nice to think so, but not very sensible. Take, for example, the decisions on nuclear testing. It is clear that the scientific-military part of the argument will not be conducted in public, either in the United States or anywhere else. And it is that part of the argument which, among the pressures and in the typical conditions of closed politics, is certain to prevail.

As for open politics, in any parliamentary democracy, just as much as in a communist society, it would be easy for any government, once the secret decision is taken, to get it not only accepted but applauded. It would be true, whatever the nature of the decision. A government has only to state any of the following propositions:

1. Nuclear testing is essential for the country's military strength;
2. Nuclear testing will make a perceptible addition to the country's military strength, enough to outweigh other considerations;
3. Nuclear testing will make a perceptible addition to the country's military strength, but not enough to outweigh other considerations;
4. Nuclear testing will make a negligible addition to the country's military strength.

None of these propositions can possibly be examined, even by scientifically trained citizens, unless they have access, which by definition they can't have, to the secret arguments; while for most citizens, including the highly intelligent, the propositions could not be examined, even if the secret arguments were thrown open. Because, while secrecy is a major obstacle, the difficulty of communicating scientific arguments is nearly as absolute. Throughout the whole of these discussions of the past year, about strategic bombing, priority for radar, Tizard and Lindemann, I have been more than ever impressed by this gap in communication. Some of the most intelligent people in the world cannot really comprehend the nature—and the fallibility—of scientific judgment. So that, over the hard core of the argument on nuclear testing, just as over the major scientific arguments of the 1939–45 war, politicians, responsible in form for the final decision, have in the long run to trust their scientists. Of course, it is not always easy to know which scientists to trust.

This is still our condition. There are some palliatives, but for the present, by which I mean the next ten or twenty years, there is no root-and-branch solution. In the long run we are forced to depend, much more than is healthy for a society, on scientific judgment. That is, the scientific judgment of a comparatively small number of men. By "scientific judgment" here I mean *scientific judgment as applied to practical affairs, and especially to military affairs.* I don't mean scientific judgment *tout court,* which is one of the greatest gifts a pure scientist can possess, the gift which tells him the problems worth doing and the problems which will "go." Rutherford had this gift to a superlative degree. So did Fermi. So did most, though not quite all, of the major scientific figures of this century. It is sometimes, but not invariably, combined with scientific judgment in the more mundane sense. It is this second kind of scientific judgment which governs our major decisions. It is, often without our knowing it, the heart of our problem. In the rest of this appendix, when I use the phrase, I shall, unless I specify otherwise, be intending it in this sense, scientific judgment as applied to practical (often military) affairs.

Scientific judgment has a good many of the properties of ordinary human judgment, judgment as we talk about it in everyday speech. It is a kind of informed and experienced guessing. It is not a "logical" process, or at least only in part. It is not necessarily connected with the highest creative talent in science, or with the most sparkling scientific or intellectual brilliance. It can be, but it need not be. It is not a romantic gift. It consists very largely in having a feel for limits, of sensing what brute nature will or will not do, of having a nose for what is "on."

Again, as with ordinary human judgment, no one possesses this gift completely or all the time. That is, everyone is fallible: but some are a good deal more fallible than others.

There is one difference from ordinary human judgment. It is not easy, in either case, to predict in advance whether a man possesses the quality, With scientific judgment, however, it is usually possible to test the result after the event.

Since everyone is fallible, no one's scientific judgment ought to be relied on uniquely and by itself. That is one simple lesson to which I shall return. But it is also imperative to identify men of good scientific judgment and get them near the decisions. We now know enough at least to be suspicious of bad scientific judgment, and to some extent of the kinds of personal-scientific temperament which are most likely to possess it. This is negative guidance, but it is better than nothing. Just realising the dangers is in itself a good deal better than nothing.

IV.
Tizard

The picture of Tizard that I gave and which others[2] had given before me has not been seriously challenged. Even those who disagree with me on other points[3] have gone out of their way to pay the highest tributes to Tizard and his achievements. Both in print and in private letters there has been a rush of praise. Praise which has that special tone and quality which only happens when people are genuinely relieved to give it. There has been a solitary criticism, but a surprising one. Lord Birkenhead in his official biography of Lord Cherwell, *The Professor and the Prime Minister,* has suggested[4] that Tizard was resistant to new ideas, unless they were his own. This is the opposite of the truth. It has been contradicted by those[5] who have occupied the highest places in English governmental science and who saw Tizard in action over a period of twenty years.

In fact, Tizard did not have much opinion of his own ideas. He was not an inventive scientist at all, in that sense. His supreme gift lay in judging the ideas of others, of knowing which to back and which not to back, and, above all, of being able to introduce the workable ones into military operations. Scientific ideas are no good in war unless they can be made into hardware with which soldiers can be taught to fight. It was this insight, both intellectual and intuitive, into actual military operations, that made Tizard so valuable in the science of war.

There is, fortunately, no doubt about his value. It is on the record, and cannot be argued with. To him, as much as to any single man, we owed the fact that there was an adequate fighter defence in 1940—only just adequate, but still it dragged us through. Of this fighter system, radar was only one part, though an essential one. Tizard's was the strongest scientific mind behind the whole system. He was at one throughout with the serving officers. This mutual trust enabled him and his committee to introduce the operational use of radar just in time.

The introduction of radar into Fighter Command between 1937 and 1940 is a textbook example of the successful application of science to war. It ought to be studied in staff colleges: and it ought to be studied by anyone who still thinks that scientific war is an affair of bright ideas. Radar was invented—that is, wireless waves were shown to be echoed from aircraft—almost simultaneously by government scientists in the United States, Great Britain, and Germany, entirely unknown to each other. As a matter of history, two Americans in the Navy Department got in first. But the British had two advantages. First, the British inventor, Watson Watt, believed passionately and imaginatively in what he was doing. Second, Tizard right from the start realised that this was the only scientific defence worth looking at, and had the power and the military insight to teach the Air Force so. *Tizard actually persuaded the Air Force to base their defensive planning on the assumption that radar would work* long before the stations existed as practical systems. This was an act of astonishing intellectual courage. Not only Tizard deserves the highest credit for it, but also the officers of Fighter Command.

By 1940 German radar was technically somewhat more sophisticated than the British: but the combination of Tizard and the officers of Fighter Command had ensured that the British knew better how to use it as a weapon of war.

That was Tizard's greatest service to the country. He did another, almost equally important, of a not dissimilar kind. Watson Watt reports[6] that as early as February 1936 he (Watson Watt) was pressing for the development of short-wave radar. It was obvious that, if one could produce a powerful source of centimetric waves, the efficacy of radar would increase—almost out of belief, by the standards of 1936. But the invention of such a source became one of the most difficult problems of the entire scientific war. Tizard, so Cockcroft has told us,[7] got the Cavendish scientists thinking about it by 1938. At some date early in the war[8] Lindemann was setting his own team on to the problem. Over this development there was no disagreement, and Lindemann, when he had full power, threw it behind centimetric techniques.

Almost everyone, however, was concentrating on a device called the klystron, which, though useful, was not the final answer. Without anyone really knowing, an approach to the final answer had been set in motion earlier, Tizard had not only talked to the Cavendish, but to Oliphant's laboratory at Birmingham. In August 1939 a number of Oliphant's staff paid a visit of several weeks to the chain station at Ventnor. In an interview with the Australian periodical *Meanjin Quarterly* Sir Mark Oliphant has stated:

> I must say that I agree completely with Blackett's appraisement of the "history" of the Lindemann-Tizard controversies. I was a colleague of Blackett's in Cambridge from 1927 to 1933 and he has been a close friend since then. I knew Tizard well, from about 1932 until

his death, and was fortunate to become his friend. Lindemann and I became closely associated from about 1937, although I knew him much earlier. However, although our relationship was always friendly, I cannot claim to have become close enough to him personally to call him other than a professional colleague.

When war broke out I had been associated, as a nuclear scientist, with the appraisal of the possibilities of using nuclear fission to produce a powerful military weapon, and some work was going on in my laboratory. However, through the vision of Tizard and Blackett, many of us who were physicists had been introduced to the concept of radar and had spent some weeks at the newly created chain of radar stations along the eastern and southern coasts of Britain. I was then Professor of Physics in Birmingham, and in view of the national situation I agreed to drop the work on atomic energy and to devote the efforts of my team to the further development of radar. We undertook, in particular, to try to develop generators and detectors of radio waves of far greater frequency than was then available.

J. T. Randall's account in a paper compiled at the request of S.R.E. Department, Admiralty, 1943, *Development of the Multi-Resonator Magnetron in the University of Birmingham,* gives much more detail.

The Birmingham scientists took over the centimetric problem. Even here nearly everyone was trying to make improvements to the klystron. Randall and Boot, in Randall's own words,[9] were standing on the side lines. Suddenly the idea broke through. Within a time that still seems magically short, Randall and Boot had a cavity magnetron working. The idea of the magnetron and its design were on paper in November 1939, and it operated in its crude form (with sealing-wax joints and so on) before the end of February 1940.

It was a device—if one is going to use such words—of genius. James Phinney Baxter was writing of the cavity magnetron when, in describing the Tizard-Crockcroft mission to the United States in 1940, he said it was the most valuable cargo ever brought to American shores.[10] Of all the scientific inventions in the Hitler war, the cavity magnetron was the most valuable.

Tizard's contribution to the scientific war is, as I have said, on the record. It is not ambiguous, and we can leave it there.

Now I should like to introduce two reflections about it, on my own account. The first has arisen by implication in some of the discussions. The second has come to me as I have thought over the whole story.

In my original lectures, I said that Tizard's was the best scientific mind that in England had ever applied itself to war. I knew precisely what I meant: but the phrase may have conveyed something slightly different from what I intended. I should now use instead of "best" the same adjective that I have just applied to the magnetron. I should write that Tizard's was the "most valuable scientific mind that in England had ever applied itself to war."

I did not mean, though I now see that I could be taken to mean, that he was the best scientist, *as a scientist,* who happened to have had a share in military decisions. That would be quite untrue. Tizard, who was very modest about his scientific attainments, would have said trenchantly that it was untrue. Take the English physicists who were active in the 1939–45 war, Blackett, Chadwick, Cockcroft, Appleton, Thomson: all Nobel Prize winners and scientists of the highest class. Tizard knew that he did not fight at the same scientific weight. He knew also that he did not fight at the same scientific weight as Oliphant, Bullard, Dee, Bernal, and a good many others. I am fairly sure, though I don't remember discussing it with him, that he would have thought Lindemann, as a pure scientist, markedly better than himself.

It is true that, just at the end of the war, Tizard's name was seriously mentioned as the next President of the Royal Society, the highest official honour that English science can give. But he took it to be right that ultimately the election went to Robert Robinson, one of the most distinguished organic chemists in the world.

Tizard was not, and probably could never have been, one of the great pure scientists. On the other hand, you would have to go a long way to find his equal for scientific judgment applied to practical affairs. That was his gift. It was because of it that he was able to make his major contribution.

The second point which I have considered is a quality of Tizard's own nature. The description I gave of him is, I think, all right as far as it goes. But I couldn't dig very deep within the limits of this kind of essay, and Tizard's was one of those characters which got more complex, more intricate in structure, the further one got into it. That was why, on page 9, I mentioned that he might draw the attention of the mature novelist. The contrast between the controlled and disciplined behaviour of the high scientist-administrator and the inner nature which he had to control and discipline would be a fascinating fob. But Tizard, it now seems to me, paid a price for this self-discipline, heavier than most of us realise. I suspect that, quite apart from the inner strains, he sometimes paid a price in action.

He had adapted himself to play the game according to the rules, the rules of the English governing and administrative classes. Like a good many deep and turbulent natures who adapt themselves so, he did it a shade too thoroughly. For most of his official life, this did not affect him; but when, as in the strategic bombing controversy, he found himself in a situation not played according to the rules, he was at a loss.

The result was, he surrendered a little easily. A less disciplined character might have made more of a row, kicked over the applecart, risked a public scandal—there were ways of doing this, difficult and painful but still possible, even in wartime. The younger scientists could not have done it with effect, but Tizard might.

For him, however, it was psychologically inconceivable. He had made a refuge for himself—and to an extent from himself. The refuge was precious to him. It was the good opinion, the company, fellowship, and support of the scientists of the Royal Society, the serving officers, the administrators, the Athenaeum. In the long run, he couldn't break away. Any action that they would condemn, even though they sympathised with him, he couldn't manage. Anything they couldn't do, he couldn't do either.

V.
Lindemann

I am going to begin with my one modest grumble. Quite a number of critics—most of them not familiar with these events and not committed to either side in any of these controversies—have scolded me, good-temperedly and charitably, for setting the issue in terms too black and white. They conclude—and in many cases this has been intended as an excuse for me—that the imagination of a novelist had been too strong. I had succumbed to the hypnosis of my own technique, they said with good nature: I had written this anecdote as though it had occurred in one of my own novels. This has been said by people who know and understand my novels very well.[11]

When they said it, though, they must have forgotten what my novels are really like. Take this situation. A is a man almost universally popular with his professional colleagues; B is, on the contrary, very widely distrusted and disliked. They are put in the middle of major decisions. A turns out to be almost invariably right; B is, on the contrary, singularly and often perversely wrong. I ask mildly: does that sound like a situation from any piece of fiction that I have written or am ever likely to write? In fiction, of course, I should have played every kind of finesse. By the time I had finished you wouldn't have known which was liked and which was disliked: you certainly wouldn't have known which was right and which was wrong.

Lord Birkenhead[12] complains that I have gone in for melodrama. Well. Have another look at A and B. Once they were bosom friends. They were associated in great hazards in the 1914–18 war. Afterwards they advanced each other's careers, were allies at the university, were linked by family life. Suddenly, without A's knowing it, B's feeling changes. He begins to talk of A as "that wretched little man." They are brought together as members of a committee, in a task on which the safety of the country may depend. B abuses A, in front of the committee, in terms of uncontrolled violence. Other members force B off the committee. A has a period of great influence. Again a twist of fortune: B is given the authority. The time is the summer of 1940, the country is in danger, A sends a message through a common friend: can't they patch up the quarrel and work together? B returns the answer: "Now that I am in a position of power, a lot of my old friends have come sniffing round."

Melodrama? Certainly. But whose melodrama? The story comes direct from Lord Birkenhead's official biography.

I could never have invented anything like that. But life is sometimes more grotesque than art. It is not sensible to talk about a novelist's temptations, when those are temptations no decent novelist would ever feel, much less succumb to.

That is my only personal grumble. Having uttered it, I

have no other complaints to make. Let us try to get back
to the plane of reason. We ought to remember from the
start that, except for the authors of *The Strategic Air Offensive,* none of the persons who have taken part in these
discussions are, in any sense, trained historians. So far as
I know, only one eminent professional historian has in-
spected the most recent evidence, Professor J. H. Plumb.
He has pointed out[13] the inadequacies in method in Lord
Birkenhead's handling of his material, and the questions
a professional historian would ask. I am sure that similar
criticisms apply to me. No doubt the historians will come
along in time. Meanwhile, the best we can do is to go
ahead in our own amateur fashion, and see if we cannot
get a bit nearer to the truth.

Lindemann's personality is likely to puzzle historians,
when they look into these events. We now know a good
deal about his *behaviour,* not only in his official, but in his
private life. For instance, he quarrelled with his sister because she married a man he disapproved of, and didn't
speak to her for forty years.[14] He cut his younger brother
just as completely out of his life.[15] After boyhood, he was
on bad terms with his mother.[16] In all these intimate relations, he showed a life pattern which might have come
out of a psychiatrist's case book. I am very chary of shorthand explanations, but I do not think it is fanciful to
pick out the same life pattern in his behaviour to Tizard.

Yet, though there was so much in his behaviour that no
one in his senses could defend, those who liked him best
insist—and this I respect—that there was a kind of essential Lindemann, different in kind from much that he did

or said. As I wrote on page 9 of *Science and Government,* he was a heavyweight of personality. He proved himself, beyond any possibility of argument, dangerous and vindictive in action, outside the ordinary human run. At the same time, he inspired passionately protective feelings in many,[17] as though he were a man longing for happiness, not knowing what was wrong with him, crying inarticulately for help. He struck his friends with the pathos of an intensely passionate man, lacking in self-knowledge, constantly struggling with a major repression or conflict which he could neither dislodge nor resolve.

Detached observers, reading these various accounts of him, will form their own judgment, according to their interpretations of personality and their human values. In his summing up, Lord Birkenhead[18] speaks of "the fundamental simplicity of his nature." Paradoxically, to me that sounds right. His behaviour was bizarre, but the structure of his nature, unlike Tizard's, may not have been complicated at all.

My essential concern, in telling this parable or anecdote, was to draw a practical lesson. From that point of view, it doesn't matter that people will form different opinions of what Lindemann was really like. Of course, it is fascinating to speculate about his character if one is interested in the human variety. It is fascinating, but it is not the point.

I should be happy if anyone who wanted to follow my practical argument read the two biographical studies of Lindemann,[19] made his own picture of the man, and read it into my story. For myself, I can accept much of the psychological portraiture of both these books. (Harrod in particular has a depth of insight any novelist would be proud of.) It is not Lindemann's character which is vital to this story.

In the same way, we need not spend much time on his accomplishments as a pure scientist. These are, in any case, less mysterious than human character, and more susceptible to objective evaluation. Scientists are judged, not by what their friends say about them, nor their enemies either, but by what they have done. Lindemann's publications are in print in the scientific journals. Anyone who wants to do so can inspect them, and any physicist can form an opinion of how important his achievement was. The language of science is international. The opinions formed by, say, Japanese or Swedish physicists would be near enough the same. I will return in a moment to what they would be likely to be.

Meanwhile I should like to mention two contributions of Lindemann's which no one has ever disputed or would wish to dispute. Neither of them was in the narrow sense scientific, but the second had big scientific consequences. I did not mention these contributions in the Godkin Lectures, except for one passing reference, on page 26. I have since regretted this. Neither was strictly relevant to my narrative, but leaving them out made the account more grudging in spirit than it should have been.

The first was Lindemann's recognition of the danger of Hitler right from 1933. This recognition was, of course, common enough in radical intellectual circles; but it was fairly rare among Tories, and to have it proclaimed, with a characteristic intensity, by Lindemann, a man far out on the right, did great good.

The second contribution followed from the first. Lindemann, believing that Hitler meant what he said about the Jews, went round Germany rescuing some of the best Jewish scientists. It didn't matter that Lindemann himself professed a rather silly anti-Semitism.[20] He brought some of the best scientific minds in Europe—Francis Simon, Kurti, the Londons, Mendelssohn, Kuhn—to Oxford. Lindemann had already done something by 1933 to improve the reputation of the Clarendon Laboratory, but it was still pretty low. The introduction of Simon and his colleagues meant that within twenty years the Clarendon was a laboratory among the top three or four in England, and perhaps in some respects the best. This must be put to Lindemann's permanent credit.

Others have made great physics laboratories. In almost every case they have been men who themselves were scientists in the full radiance of their own creative work. This was true of Rutherford at the Cavendish, Franck at Göttingen, Bohr at Copenhagen, Ernest Lawrence at Berkeley. I can only think of two exceptions. Between 1922 and 1938 Rome suddenly became one of the great physical centres of the world, and produced masters like Fermi, Segré, Rasetti, Amaldi. But these have all told us that the origin of this efflorescence was none of them, but a father-figure called Corbino, senator, banker, impresario, man of affairs, who had, in the intervals of making a fortune, presided over the laboratory, found the money, collected the talent. He brought together a great laboratory until

Fascism destroyed it. Lindemann's achievement with the Clarendon was not dissimilar in kind.

As for his own original achievement in pure science, that—if it were not for the fact that he was, by common consent, both an exceptionally clever man and a controversial figure—does not leave room for genuine disagreement. Studying his publications, out of range of the aura of his personality, our putative Swedish and Japanese referees would report that:

(*a*) Between 1910 and 1914 he produced nearly twenty papers. There are signs of overmuch diversion of effort, but some of these papers are original and deep.

(*b*) After 1919 he produced a number of notes and letters on many subjects, but little substantial work; although he was working at a period when physics, both theoretical and experimental, was wide open to discovery as it may not be again for generations, he did not find a profitable field. None of this would surprise the referees very much. There are many physicists and mathematicians who make a good start and then find their creative impulse dry up very early. Some of them undergo bitter personal suffering on this account.

But Lindemann's *oeuvre* as a pure scientist does not bear much on the problems of his public career, any more than his personality does. If he had had the personality of a Schweitzer and the scientific *oeuvre* of a Rutherford or a Bohr, the problems would still remain. The discussion, at its core, is not concerned either with personality or with creative work. It is concerned with scientific judgment.

VI.

It was in the struggles with Tizard that Lindemann's scientific judgment was called most violently into question —though there were similar controversies where Tizard was not concerned, which, as the histories are written, we shall hear more about in the next ten years. One or two of these I shall mention in passing. For the present, it is enough to see what more we can learn about the struggles with Tizard.

The Personal Story

There is not much to add; and what there is (compare the last paragraph of Section V) does not really affect the central problem. Possibly no one will ever know when and why the personal antagonism blew up. It does not seem quite so surprising, now that we have learned of Lindemann's unilateral breaches inside his family. It certainly seemed surprising to Tizard, however. There have been attempts to represent it as a straight struggle for power: but that is an oversimplification. On Lindemann's side the hatred went on till death. There was nothing like the same emotional involvement on the part of Tizard. To say he felt specially benevolent to Lindemann would be sentimental, but, in his last years, he spoke of Lindemann and the whole story with a kind of sardonic and lively detachment. He went out of his way to reiterate[21] that Lindemann was one of the cleverest men he had ever met.

The Tizard Committee and Radar

A lot of people have become baffled and lost by what has seemed a conflict of evidence. Sir Robert Watson Watt[22] has gone on record that Lindemann encouraged his radar work and gave him active support in 1936 during the period when Lindemann was sitting on the Tizard Committee. Professor R. V. Jones says[23] that he remembers discussing with him in 1935 the possibilities of radar. Both these are men of complete integrity.

On the other hand, there were five people who were witnesses of all that happened on the Tizard Committee throughout its life—Tizard himself, Wimperis, Blackett, Hill, and Rowe. Of these five, Tizard and Wimperis are dead. Before he died, Tizard gave an account of Lindemann's behaviour on the Committee which was essentially identical with mine on pages 26-31 of *Science and Government.* Tizard gave this account, not in ordinary conversation, though he did that too, but in reply to questions by an official historian.[24] Wimperis in conversation gave the same account. So did Blackett in print, a year before my lecture was delivered, and again in confirmation of what I had said.[25] Rowe, who was secretary of the Committee throughout, has publicly confirmed my account.[26] Hill did so privately while it was in manuscript,[27] and had himself, though this has generally not been realised, published the only first-hand description of the proceedings actually written at the time.[28]

The integrity of all these men, just as with Watson Watt and Jones, is complete.

It might seem that the collision is head-on. Who is wrong? Whose memory is defective? Where is the contradiction?

The answer is undramatic and something of an anticlimax. No one is wrong. No one's memory is defective. There is no contradiction.

The confusion has arisen very largely because most people are not familiar with the nature of scientificmilitary arguments. The technical dispute on the Tizard Committee, like eight out of ten scientific-military arguments, was about priorities. Priorities mean what they say. They are arrangements in order of importance. Given infinite resources, in scientific war, one would often like to try almost everything—as the Americans did in making the fusion bomb. But usually that is not possible, even in America. One has to establish priorities, to see that the best chance is not missed, even if it means sacrificing lesser chances.

Stripped of its violent personal emotion, the issue before the Committee was a judgment of priorities. The technical disagreement was about that, and nothing else. It is now almost certain that before the Committee ever met Tizard had decided that radar had to be given something like absolute priority. Sir Philip Joubert[29] remembers him late in 1934 dismissing the possibilities of air defence one after the other—"Infrared. No *real* good," and so on—then telling the officers that he was pretty sure they had got the answer: and, even as early as that, beginning to talk in terms of military use. He never budged from this position. He wanted as near absolute priority for radar as he could get, not only in technical development but, at least as imperatively, in terms of indoctrinating the services. It is very rare, of course, to be able to get absolute priority in the literal sense for any scientific-military scheme. Tizard and the Committee were prepared to expend a little effort on (that is, give a low but finite priority to) other schemes such as infrared detection, though it seemed highly improbable it could be any use in time for a war in 1939 or 1940. But they were not ready to make many compromises about priority, and in fact never did so.

As soon as Lindemann took his place on the Committee, he wanted to alter Tizard's order of priorities. He was actively interested in Watson Watt's work, but he was not prepared to give radar the near-absolute priority that the Committee had already settled on. The Committee's preoccupation—its all-important preoccupation—with the operational use of radar meant nothing to him. He had his own order of priorities. This order appears to have varied from time to time. He had a number of his own specifics—wire-carrying shells, aerial mines carried by parachutes, bombs carried by wires, as well as infrared detection—which he was determined to press on to the Committee. His favourite devices seem to have been bombs dangled by wires in front of enemy aircraft and the parachute-carried aerial mines. Sometimes he argued that the parachute mine deserved higher priority than radar. At no time was he prepared to accept it as being less than equally important or as carrying less than equal priority.

Lindemann's general idea of appropriate priorities—about which the evidence of all present is unanimous—we can, fortunately, confirm by the aid of two documents written at the time. We shall only know the precise details when the minutes of the Committee are made public. It seems a pity that these minutes, which were kept by Rowe, are not already available. They would say the last word on an interesting piece of scientific-military history. However, we have two documents written in 1936. The first was composed in verse by A. V. Hill, shortly before the Committee was coming to its final arguments. It is published in his book *The Ethical Dilemma of Science,* 1959, with the following prefatory note:

> The following poem, in the style of the Earl of Derby's translation of the Iliad (1864), purports to represent the Minutes of a meeting of a Committee of the Air Ministry in 1936, together with a summons to the next one. These meetings were *secret,* and even to-day, twenty-three years later, considerations of propriety, if not of security, require that pseudonyms should be used: this may explain how a Norse deity and a Geheimrat somehow got mixed up with a lot of Greek characters on a Trojan Committee.*

---

* Characters in the poem in order of appearance:

Sigma	Tizard
Omega	Wimperis
Theta	Blackett
von Alpha-plus	Lindemann
Phi	A. V. Hill
Odin	Sir Winston Churchill

Attending there on ancient Sigma sat
The Elders of the City: Omega
And Theta and von Alpha-plus and Phi.
All these were gathered at Adastral House,
By age exempt from war, but in discourse
Abundant as the cricket that on high
From topmost bough of forest tree sends forth
His music: so they sent their Minutes forth,
And all men wondered, even Odin wept
With tears of joy that Ilium was safe.

Von Alpha-plus arose and thus began,
"O ancient Sigma eminent in war
And in the council wise: thy present words
No Trojan can gainsay, and yet the end
Thou hast not reached, the object of debate.
This city cannot be immune from war
Until a hail of parachuting mines
Descend unceasing at its eastern gate.
So shall the long-haired Greeks remain at home
Nor lay their infernal eggs upon our streets."

Thus angrily, and round his body flung
His cloak, and on his head a billycock,
Then passing cocked a snook at Lambda-Mu,

Lambda-Mu	Roxbee Cox	who attended some
Hermes	Joubert	meetings of the Committee
Rho	A. P. Rowe	
Hopskipjump	Lord Caldecote (at this time	
	Sir Thomas Inskip)	

SCIENCE AND GOVERNMENT

103

Last called his shiny Rolls of eighty steeds
And soon without the tent of Odin stood.
Him, from his godlike sleep, he sought to rouse
Loud shouting: soon his voice his senses reached:
Forth in his slumber-suit bearlike he came
And spoke to deep designing Alpha-plus,

"What cause so urgent leads you through the camp,
In the dark night to wander thus alone?"

To whom von Alpha-plus of deep design replied,
"O Odin, godlike son of destiny, awake:
For ancient Sigma's professorial crew,
With Hermes of the glancing wings and Rho
Who keeps the minutes but who wastes the hours,
Will not be happy till the long-haired Greeks
Upon this city lay their infernal eggs.
They have no mind to fill the sky with mines
Attached to parachutes: and precious days they waste
In vain experiment with R.D.F.
If, godlike son of destiny, we two
In place of Hopskipjump and Sigma were
Thy sky would rain with parachuting mines
Unceasing, and the land be safe." So spake
Von Alpha-plus of deep and bold design.
Him answering, Odin, son of destiny, replied,
"Many indeed, and fierce, the bombs I've dropped,
But never 2-oz. mines attached by wires
To parachutes, by day and night alike,
In billions at our eastern gate. The like
Has never been before. We two will take

The tidings to the Minister of State.

    With Odin Lord Almighty of land and sky and sea
    And Alpha-plus to help him, how happy all will be!"

So ancient Sigma and his stag-eyed crew,
Theta with bright ideas, Phi with none,
Rho with the Minutes, weary Omega,
Sat long and silent in the deepening gloom,
While Lambda-Mu went out and hanged himself,
Snook-cocked by Alpha-plus of deep design.
At last with downcast visage Sigma spoke:
"The game is up. Without von Alpha-plus,
Of wily counsel and of deep design,
Who speaks with politicians and the Press,
And soon may be M.P. for Oxenbridge,
All hope is gone and many-murdering
Death Will hunt his victims in our streets."
To which Theta of bright ideas, Phi of none,
Rho of the Minutes, weary Omega,
Had nothing printable to add. But set
A day to meet Geheimrat Alpha-plus
And pray for mercy from his mighty friends,
From Odin, godlike son of destiny,
And from himself, the man of deep design.
Then ancient Sigma and his stag-eyed crew
Will make submission to von Alpha-plus
(Except for Lambda-Mu who hanged himself).
Your presence is requested at 11:
The number of the room is 008.

The second document—and this is the most satisfac-
tory kind of evidence—was written by Lindemann him-

self. In a letter to Churchill dated February 27, 1936, he said:

> The only part of the Committee's work which has so far been successful has been the development of methods of detection and location. The reason for this seems to be that it has been put in the hands of a man who suggested the method and believed in it and that he could and did push ahead with whatever experiments he thought necessary.
>
> I suggest however that the only way of making progress with the equally important development of aerial mines and the related question of shellburst which remain effective for some reasonable period, is to put them in the hands of some enthusiastic believer who is not compelled to come back to the Committee every time he wants to make a fresh experiment.

It was the "equally important development of aerial mines" with which he had worn the Committee down. Anyone who could suggest giving this device equal priority with radar seemed to them so lacking in scientific judgment as to be a danger. Scientists are used to harsh argument, but they were not prepared to argue indefinitely about aerial mines and wire-carried bombs. These ideas made no technical sense. Time was being wasted, and wasted in an atmosphere of extreme hostility. The only sane course was to get back to the highest priority for radar and for the operational use of radar.

There has been some speculation as to what—on the level of reason and purpose—Lindemann was really after. One view, which Rowe is inclined to hold,[30] is that he was determined to wreck the Committee at any cost, using any scientific idea regardless of its merit, so that in the long run he could get the power himself. This, of course, would be irresponsible behaviour by any standards, since this Committee was the only body in the country working out a scientific air defence, and war was very near.

The second view is that simply in advancing these fantasy projects, in haranguing the Committee about them, he really believed what he said. This view is entirely consistent with some of his behaviour later in his career. It is also consistent with the fact that, after he had been forced off the Committee, he was still pressing the wire-carrying shell and the aerial-mine parachute on to various Government departments.[31]

If I were writing my account of the Tizard Committee again, I should include some of what I have just said, but otherwise alter nothing. I should underline the question of priorities, though in fact in my original account I chose my words with care (see pages 29–30 of *Science and Government* where the issue of priorities is precisely stated). I should include the two documents I have just quoted. They make the whole dispute clear.

There is one omission I should now make. I should leave out the last two paragraphs of Section VI, pages 32–33. I should not have permitted myself to go in for such a speculation. I should now finish the story of the Tizard Committee by saying something plain and prosaic. It is now clear that any body, setting up a scientific air defence in Great Britain in 1935, had to do three things: (1) it had to give high priority to radar; (2) it had to make an act of faith, and persuade the serving officers of the use of radar before the apparatus was so much as in existence: it had to go on concentrating on the operational use of radar; (3) it had to let university physicists into the secret, so as to get them thinking about centimetric radar. No other policy would have given us much chance of surviving. Fortunately for us, this was the policy of Tizard and his Committee.

Strategic Bombing

Over the 1942 controversy on the effects of strategic bombing, the facts are more eloquent than any lucubrations about them. These facts, which are now known in detail, are almost exactly as I stated them. Lindemann's scientific judgment was wrong. He overestimated the facts of bombing by at least ten times. Tizard's judgment was much more nearly accurate. Blackett's was better still. The only documents of importance which remain to be examined are Blackett's memoranda, still hidden in the Admiralty files. Since Blackett, partly for interdepartmental reasons, took the argument considerably further than Tizard did, his papers are needed to complete the story. We have, however, his published recollections of these events.

The facts are so eloquent that, if I had had the Lindemann and Tizard papers before me, I should just have printed them without comment. Let me quote the official history (the footnotes are part of the quotation):

It was at this moment of the crisis of strategic bombing that Lord Cherwell intervened. On 30th March

1942, he addressed a minute to the Prime Minister in which he said:

"The following seems a simple method of estimating what we could do by bombing Germany:

"Careful analysis of the effects of raids on Birmingham, Hull and elsewhere have shown that, on the average, 1 ton of bombs dropped on a built-up area demolishes 20–40 dwellings and turns 100—200 people out of house and home.

"We know from our experience that we can count on nearly 14 operational sorties per bomber produced. The average lift of the bombers we are going to produce over the next 15 months will be about 3 tons. It follows that each of these bombers will in its lifetime drop about 40 tons of bombs. If these are dropped on built-up areas they will make 4,000–8,000 people homeless.

"In 1938 over 22 million Germans lived in 58 towns of over 100,000 inhabitants, which, with modern equipment, should be easy to find and hit. Our forecast output of heavy bombers (including Wellingtons) between now and the middle of 1943 is about 10,000. If even half the total load of 10,000 bombers were dropped on the built-up areas of these 58 German towns the great majority of their inhabitants (about one-third of the German population) would be turned out of house and home.

"Investigation seems to show that having one's house demolished is most damaging to morale. People seem to mind it more than having their friend or even relatives killed. At Hull signs of strain were evident, though only one-tenth of the houses were demolished. On the above figures we should be able to do ten times as much harm to each of the 58 principal German towns. There seems little doubt that this would break the spirit of the people.

"Our calculation assumes, of course, that we really get one-half of our bombs into built-up areas. On the other hand, no account is taken of the large promised American production (6,000 heavy bombers in the period in question). Nor has regard been paid to the inevitable damage to factories, communications etc. in these towns and the damage by fire, probably accentuated by breakdown of public services."*

. . . The Cherwell minute, therefore, involved certain matters of high policy such as the production programme for heavy bombers and the concentration of the bombing offensive upon certain strategic objectives. It also involved calculations of probability such as the average life of an operational bomber in 1942 and 1943 and the effect of a certain weight of bombs upon a given number of German towns. Clearly the

* Min. Cherwell to Churchill, 30th March 1942. Circulated by the Prime Minister to Defence Committee on 9th April 1942. The whole text of the minute is transcribed above.

results which Lord Cherwell foresaw in consequence of his probability calculations would not be achieved unless policy decisions were taken in the sense which his minute suggested as necessary. On the other hand, these policy decisions could not be justified by the probability calculations themselves because the calculations seemed probable only to those who, in any case, believed in the policy. To those who did not, they seemed to be wholly improbable and, in this connection, one scientist could easily be answered, or at least questioned, by another. Sir Henry Tizard, for example, observed that "the risk entailed by this policy is so great that it is necessary to be convinced not merely that it has a chance of success but that the probability of success is very great."*

To Sir Henry Tizard it seemed that Lord Cherwell's calculations contained certain important fallacies. He pointed out that the Ministry of Aircraft Production programme provided for the construction of 3,585 Wellingtons and 5,219 heavy bombers between the beginning of April 1942 and the end of June 1943. 689 of the Wellingtons were earmarked for Coastal Command. Thus, the bomber programme provided for 8,115 aircraft. Experience had taught Sir Henry Tizard "that we cannot rely on more than 85% of the target programme" and he therefore estimated that Bomber Command would receive 7,000 and not 10,000 aircraft in the period reviewed by Lord Cherwell.

Even if this difficulty could be overcome, Sir Henry

---

* Memo. by Tizard, 20th April 1942.

Tizard immediately saw another in the assumption that each of these aircraft would on the average complete fourteen operational sorties. This would mean that they would all be destroyed and that "we should be left at the end of the period with a front line strength no greater than it is at present, which is surely quite unthinkable." The two difficulties led Sir Henry Tizard to the conclusion that Bomber Command would be able to drop on Germany only half the tonnage which had been estimated by Lord Cherwell.

The next point which struck Sir Henry Tizard as "much too optimistic" was the assumption that the "58 towns of over 100,000 inhabitants" would be easy to find and hit. He thought that Lord Cherwell had underestimated the difficulties which would confront the bomber crews operating at night in the face of heavy opposition. *Gee,* he pointed out, had a limited range and would have a limited life. New radar aids were not expected by Sir Henry Tizard to come into service until April 1943.* He, therefore, thought it unsafe to assume that more than twenty-five per cent of the bombs lifted would find their targets. Thus, he calculated that in the period reviewed by Lord Cherwell and on the assumption that all heavy bombers were concentrated exclusively on the task,† not more than 50,000 tons of bombs would fall on the built-up areas. If this was spread over the fifty-eight towns the effect might, on

---

* In which expectation he was not far wrong.

† Which concentration, Sir Henry Tizard thought, would be neither wise nor possible.

the average, be three or four times as great as that produced by the Germans in Hull and Birmingham. This, Sir Henry Tizard thought, "would certainly be most damaging but would not," he said, "be decisive unless in the intervening period Germany was either defeated in the field by Russia, or at least prevented from any substantial further advance, e.g. to the Russian or Iranian Oilfields."

Thus, although Sir Henry Tizard had by no means fully realised the extraordinary resilience and determination with which the Germans were to meet both the bombing of their towns and the defeat of their armies in Russia, he had sounded a note of warning against the assumptions upon which the Cherwell minute was based and so against the policy into which Bomber Command was drifting. The concluding paragraph of his memorandum is worth quoting in full:

"I conclude therefore," he wrote:

"(*a*) That a policy of bombing German towns wholesale in order to destroy dwellings cannot have a decisive effect by the middle of 1943, even if all heavy bombers and the great majority of Wellingtons produced are used primarily for this purpose.

"(*b*) That such a policy can only have a decisive effect if carried out on a much bigger scale than is envisaged in [the Cherwell minute]."*

* Memo. by Tizard, 20th April 1942. Sir Henry Tizard sent his note to Sir Archibald Sinclair and Lord Cherwell.

. . . Nevertheless, as the Prime Minister also remarked, "there must be a design and theme for bring-
ing the war to a victorious end in a reasonable period.
All the more is this necessary when under modern
conditions no large-scale offensive operation can be
launched without the preparation of elaborate technical apparatus." Lord Cherwell's minute, despite its
largely and inevitably fallacious "forecasts,"* had done
no more and no less than to acknowledge a "design
and theme" for the air offensive, and Lord Cherwell
exerted a much greater influence upon the Prime Minister than did Sir Henry Tizard.

The Air Staff, as has been shown, had already de-
vised this theme towards the end of 1941 and Lord
Cherwell had added little that was new. All the same,
because of the position which he occupied and the
time at which he submitted his minute, Lord Cherwell's intervention was of great importance. It did
much to insure the concept of strategic bombing in its
hour of crisis.[32]

To which marmoreal account I should have wished to
add only one comment. The use of the word "inevitably"

---

* There were, of course, many fallacies, some less inevitable
than others, besides those remarked upon by Sir Henry Tizard.
For example, Lord Cherwell had assumed that if one ton of
bombs dropped on a built-up area made between a hundred and
two hundred people homeless, then forty tons would make four
thousand to eight thousand people homeless. This, however, did
not necessarily follow.

seems semantically a little odd. Of course, as I have said before, all scientific judgments are informed guesses. But some informed guesses turn out nearer the truth than others.

The argument about scientific judgment is clear enough. But there was also, closely connected with it, an argument about strategic judgment. Certainly Blackett, and, I think, Tizard (though in his case I have to rely on my memories of conversations at the time, which is not good evidence), had major strategic objections to the bombing policy as a whole. They also, and this is incontrovertibly well documented, had a precise strategic objection to the bombing policy of 1942–43. They, like the Naval Staff, believed that long-range aircraft should be used for antisubmarine purposes, largely, though not entirely, in the Battle of the Atlantic. If this battle was lost, the war was lost. On the other hand, a victory over the submarines in 1942 would transform the shape of the war. To Tizard and Blackett such a victory seemed technically possible, for a relatively small investment of bombers and of radar apparatus which—according to the Air Staff–Lindemann bombing doctrine—should be hoarded for mass bombing. The official history of the naval war says:

> The basic issue which had to be settled by the Cabinet was, therefore, whether, taking account of the prevailing shortage of aircraft, a balance could be struck between the accepted Allied policy of bombing Germany and Italy as heavily as possible and the urgent need to improve the protection of our convoys. One funda-

mental requirement was to estimate just how effective the bombing of Germany had already been, and also how effective it was likely to become. Lord Cherwell forecast that in 1943 bombing of built-up districts would deprive about one-third of the population of Germany of their homes, and that this might be decisive . . .

In the early summer the Admiralty's anxiety deepened. U-boat sinkings remained very high, in the Mediterranean "the situation was precarious," the Far East "was in a state of disintegration," and our ability to hold the Indian Ocean "was in balance." "Ships alone," they said, were "unable to maintain command at sea . . . a permanent and increased share in the control of sea communications had to be borne by [the] air forces." The requirements were once again analysed, and a deficiency of 800 aircraft was arrived at. But the Air Ministry still felt that "to dissipate the Royal Air Force's strength" in order to reinforce Coastal Command would be a strategic error. They held that, as the bombing of Germany gained momentum, the threat to our sea communications was bound to diminish. By reducing the weight of our bombing we might merely postpone the day when the rising curve of Allied merchant ship construction would overtake our losses. To this argument the Admiralty's reply was that, quite apart from the great value of the ships lost, every one of their cargoes was of immense importance to the nation's war effort; that there was a real danger of our war production and transport slowing down, or even coming to a stop, through failure to bring in the essential imports of food and raw materials; that losses on the present scale could not continue without the morale of the Merchant Navy suffering; and finally that unless stronger air escorts were provided the enemy's rising U-boat strength would overwhelm the defenders of our convoys.[33]

Blackett takes up the story at this time:

From my talks with Lindemann at this time [the summer of 1942] I became aware of that trait of character which Snow so well emphasizes: this was his almost fanatical belief in some particular operation or gadget to the almost total exclusion of wider considerations. Bombing to him then seemed the one and only useful operation of the war. He said to me (unfortunately I have no record of this conversation, but he probably said the same to others) that he considered any diversion of aircraft production and supply to the anti-submarine campaign, to army co-operation or even to fighter defence—in fact, to anything but bombing—as being a disastrous mistake. Lindemann even suggested that the building up of strong land forces for the projected invasion of France was wrong. Never have I encountered such fanatical belief in the efficacy of bombing.

The high priority given thereafter to everything pertaining to the bombing offensive made it very difficult to get adequate air support for the vital Battle of the Atlantic. If this had got worse there would have been no more bombing offensive for lack of fuel and bombs, and no invasion of France in 1944. I remember that during the winter of 1942 and 1943 the Admiralty had to enlist President Roosevelt's personal influence to ensure that a squadron of that admirable antisubmarine aircraft, the B-24, was allocated to Coastal Command (where they were brilliantly successful) and not, as the Air Staff wanted, sent to bomb Berlin, for which they were not very suitable. However, at the Casablanca Conference in January, 1943, a combined American and British bombing offensive was formally adopted as a major part of the British war strategy.

No part of the war effort has been so well documented as this campaign, which had as its official objective "the destruction and dislocation of the German military, industrial and economic system and the undermining of the morale of the German people to the point where their capacity for armed resistance is fatally weakened." Immediately after the war the U.S. Strategic Bombing Survey was sent to Germany to find out what had been achieved. A very strong team (which included two men who are now advisers to President Kennedy, J. K. Galbraith and Paul Nitze) produced a brilliant report, which was published in September, 1945. Without any doubt the area-bombing offensive was an expensive failure. About 500,000 German men, women and children were killed, but in the whole bombing offensive 160,000 U.S. and British airmen, the best young men of both countries, were lost. German war production went on rising steadily until it reached its peak in August, 1944. At this time the Allies were already in Paris and the Russian armies were well into Poland. German civilian morale did not crack.

Perhaps it is not surprising that the report of the Strategic Bombing Survey seems to have had a rather small circulation; it is to be found in few libraries and does not appear to have been directly available, even to some historians of the war.

If the Allied air effort had been used more intelligently, if more aircraft had been supplied for the Battle of the Atlantic and to support the land fighting in Africa and later in France, if the bombing of Germany had been carried out with the attrition of the enemy defences in mind rather than the razing of cities to the ground, I believe the war could have been won half a year or even a year earlier. The only major campaign in modern history in which the traditional military doctrine of waging war against the enemy's armed forces was abandoned for a planned attack on its civilian life was a disastrous flop. I confess to a haunting sense of personal failure, and I am sure that Tizard felt the same way. If we had only been more persuasive and had forced people to believe our simple arithmetic, if we had fought officialdom more cleverly and lobbied ministers more vigorously, might we not have changed this decision?[34]

I have nothing to add, except the sombre words of the naval historian:

For what it is worth this writer's view is that in the early spring of 1943 we had a very narrow escape from defeat in the Atlantic; and that, had we suffered such a defeat, history would have judged that the main cause had been the lack of two more squadrons of very long range aircraft for convoy escort duties.[35]

Window

There was a third and final wartime argument between Lindemann and Tizard. I did not know enough about it to mention it in my original lectures. The subject was not of the order of importance of the other two. The official history of the bombing offensive, however, deals with the controversy in terms severely critical of Lindemann.

The full story (which the official history does not give) shows Lindemann deserving of credit at an early stage, although at the decisive time his scientific judgment of military operations was once more at fault.

As early as 1937, it had been suggested that defensive radar systems could be interfered with by the very simple device of dropping from aircraft strips of paper coated with metal. Lindemann had received this idea from R. V. Jones[36] and in 1942 gave another of his friends, Derek Jackson, the power to make operational tests.[37] So that he was responsible for the inception of a really valuable device. It is at this point that the official history takes up the story.

> Experiments had shown that the dropping of metallised strips of paper in quantity produced a reaction on radar screens operating on certain frequencies and prevented accurate measurements being made on them.

These metallised strips, which subsequently came to be known as *Window,* had the great advantage of being cheap, easy and quick to produce, and it seemed that their immediate introduction would confer upon Bomber Command an important advantage in the air battle which was now joined. For these reasons the Air Staff suggested in April 1942 that the use of *Window* should be authorised at once and the Chiefs of Staff had little difficulty in agreeing to the request . . .

It was at this stage, however, that Lord Cherwell intervened with the pertinent suggestion that *Window* might also disrupt night-fighter radar interception and after a meeting which he held with Sir Archibald Sinclair and Sir Arthur Harris early in May 1942, it was decided to defer the introduction of *Window* until these possibilities had been investigated.

Thus, at a time when the Bomber Command offensive was rapidly gaining momentum in the face of heavy and increasing casualties and at a time when the German air offensive was diminishing to negligible proportions, a cardinal weapon, favouring the bombers and hindering the defences, was cast aside for more than another whole year. It was not until the night of 24th July 1943 that Bomber Command was able to use *Window* for the first time and the sensational success which it then achieved in the famous Battle of Hamburg is, indeed, a grave verdict on the many decisions to defer its earlier introduction.[38]

And the story is continued in the second volume of the official history:

Thus, one of the principal reasons for withholding *Window,* namely the danger of revealing it to the enemy, was seriously undermined from the outset and seemed to have been completely destroyed by the end of October 1942 when an Air Scientific Intelligence report indicated that it was "certain" that the Germans fully understood the *Window* principle.

This consideration of the secret being out seemed, in the view of Sir Henry Tizard, to complete the argument for the immediate introduction of *Window,* but there were others who persevered with a different view ... This attitude and the absence of Sir Henry Tizard at a meeting at the Air Ministry on November 4, 1942 left the floor to those who were preoccupied with the dangers and not the advantages of *Window,* the introduction of which was consequently again postponed.

By the end of March 1943 when the Battle of the Ruhr had begun, the case against *Window* was, however, palpably crumbling. The obvious fact that the German bomber force was an almost negligible factor and that the German fighter force was one of increasingly decisive importance was at last beginning to exert some influence upon the discussions. Indeed, it now appeared that of the losses inflicted by enemy action upon Bomber Command no less than seventy per cent were due to German night fighters. About half these losses to night fighters were, it was estimated, attributable to radar-controlled fighters which might be substantially put out of action by *Window.* Of the remaining thirty per cent of the casualties which were attributed to flak, it was thought that two-thirds were due to radar-controlled guns which similarly might be dislocated by *Window*. Thus, it seemed probable to the Air Staff that the introduction of *Window* would save from destruction no less than thirty-five per cent of the aircraft which were then being shot down by enemy action. This would amount to 1.7 per cent of the total Bomber Command sorties being despatched.

There were, of course, astonishing defects in this optimistic argument, which, for example, assumed that the dislocation of radar would result in the G.C.I.s and A.I.-equipped night fighters and the radar-controlled anti-aircraft guns achieving absolutely no results at all. Nevertheless, the argument for introducing *Window* did seem to be overwhelming especially in view of the great weakness of the bomber force remaining to Germany with which to retaliate . . .

Even this was not the end of an already sad story. The Chiefs of Staff now decided that *Window* should be further delayed until after the projected invasion of Sicily had been carried out. By the middle of June 1943, however, even Lord Cherwell was beginning to recognise "on the whole that the time is rapidly approaching when we should allow it *(Window)* to be used." Even so, he advised the Prime Minister that the introduction of *Window* should not be allowed to endanger allied plans in the Mediterranean. Meanwhile, in the strategic air offensive, British bomber losses between 1st April and 14th July 1943 amounted to 858 aircraft. German bomber losses in the same period were, it seemed, twenty-seven aircraft. It appeared to Sir Charles

Portal that if *Window* had been used in these months, Bomber Command might have saved 230 bombers and crews and the Germans might have saved sixteen.[39]

VII.

In the war Lindemann had, so Lord Birkenhead says, "power greater than that exercised by any scientist in history."[40] That is dead true. It is precisely for that reason that his record of judgment has to be examined, if we are to extract lessons which are necessary to us all. There are two primary lessons. I don't want to labour the first. Flogging dead horses is no fun for anyone. But I also don't want to leave any vestige of ambiguity. This first lesson is not to do with how nice a man Lindemann was, nor with how good a scientist he was. It is to do, inescapably, with his scientific judgment in the sense I defined it on page 4, that is, his scientific judgment applied to practical affairs, in particular to military operations. This kind of scientific judgment is not a thing which has to be left to subjective evaluation for ever. To a meaningful extent, it can be tested. What a man recommended or predicted can be measured against what actually happened. His suggestions and decisions are on paper, and as a rule we can see what came of them. Technological history, even in the confusion of war, is more precise and ruthless than any other kind of history. By these standards, Lindemann's scientific judgment was unusually bad.

The phrase "unusually bad" is not a loose one. It doesn't

mean that we are thinking of an ideal certainty of judgment which human beings never attain. It means that, compared with a large number of other scientists in both the United States and Great Britain, making judgments about the same topics at the same time, Lindemann's record is a bad one. Compared with, say, Vannevar Bush, Conant, A. H. Compton, Lawrence, Fermi, Tizard, E. J. Williams, Cockroft, Bullard, Blackett—to mention only some of the more obvious names—there is no possible argument about who made the most misjudgments, both positive and negative. It would be a constructive exercise, and quite practicable now that many of the documents are available, to draw up a list of the decisions in which these men were involved, and note their scientific judgments at the time. None of them, I think, would come out with a score of 100 per cent right: but a good many judged well a remarkable percentage of the time.

Lindemann did much worse. He was concerned with three major scientific judgments (the air defence of Great Britain, strategic bombing, atomic energy). We have seen his record about the first two. The third can be left until the history is published. He also intervened in a very large number of minor judgments. I will limit myself to a few examples. In the controversy in 1944 between the continuation of mass bombing and the Tedder-Zuckerman policy of bombing communications, he was on the wrong side. In one or two cases, such as those of *Window* and centimetric radar, his inceptive judgment was correct, and he then damaged his own credit by going quite wrong in his military judgment about their actual use. In other interventions, the documents about which must still be lurking in Government departments, notably the Ministry of Supply, he was wrong throughout. One would like, for instance, to see the papers about tank design.

People have often wondered, with genuine puzzlement, why his judgment was so bad. After all, he was an exceptionally clever man. His scientific equipment was wide and deep enough for anything he tackled. One explanation is, of course, that he tackled too much. He had the opportunity to intervene more than any scientist in the world, and he took it. That in itself showed a defect in judgment. But I believe the real defect lay deeper.

He believed, as much as any man of his time, that he could solve any problem by his own a priori thought. This is the commonest delusion of clever men with bad judgment. For anything like reasonable judgment, a man has to know when to rely on others and when to think, alone and uninfluenced, for himself. It is precisely that balance which makes for what we call judgment. Without that balance, the cleverest of men are going to make the worst of guesses.

VIII.

The first primary lesson, then, is banal. If you are going to, have a scientist in a position of isolated power, the only scientist among nonscientists, it is dangerous when he has bad judgment.

That is a lesson which shouts at one at once. But there

is a second primary lesson which doesn't shout at one quite so loud but which a lot of us feel compelled to insist upon. If you are going to have a scientist in a position of isolated power, the only scientist among nonscientists, it is dangerous *whoever he is*. This was the lesson which burnt itself in upon many during the controversies of 1939–1945: whoever he is, whether he is the wisest scientist in the world, we must never tolerate a scientific overlord again.

During what I have just written, there have been examples of scientists whose scientific judgment was very good. But even a Vannevar Bush, in solitary power among nonscientists, carries a potential danger too heavy to be risked. Incidentally, Bush would have been the first to say so.

If I were writing the lectures again, I should bring out the lesson even more emphatically than I originally did. It is the essential negative lesson of what I am trying to say. The positive lessons remain as I wrote them, or at least I have not anything constructive to add. But the negative one needs, in our present non-scientific administrations, to be written up in great block letters.

Let me finish by a terse example out of the official history of strategic bombing.[41] In the autumn of 1942 the Admiralty was still bitterly concerned about the use of bombers. On October 30th the First Sea Lord proposed that "there should be an objective scientific analysis of the effects of the bombing offensive" and that this analysis should be made by "a committee consisting of Lord Cherwell (Lindemann), Sir Henry Tizard, Professor Bernal, Dr. Cunningham, Sir Charles Darwin and Professor Blackett."

The Chief of Air Staff reacted by saying, in an undated note, that if the Chiefs of Staff really wanted further scientific advice "Lord Cherwell should be asked to give or obtain an authoritative opinion."

Those are terrifying words. Not terrifying because Lindemann was Lindemann, though that gives them an added twist. They would be just as terrifying if, for Lord Cherwell as the source of authoritative opinion, we read Bush or Compton or Tizard or any scientist whose judgment has been proved unusually good. For the terrifying thing is, not who issues this authoritative opinion, but that intelligent and high placed nonscientists should believe that it exists. *That* is the danger of having one scientist in a position of power among nonscientists. Whatever we do, it must not happen again.

NOTES

T. P. means Tizard Papers.

SCIENCE AND GOVERNMENT

1. Don K. Price, *Government and Science* (New York University Press, New York, 1954), p. 30. Much the most interesting and experienced book on the subject that I have read. Nothing written on government and science in England remotely compares with it.

2. T. P., diary, May 8, 1945.

3. T. P., autobiography, MS, p. 17.

4. House of Lords Hansard, 1957, weekly No. 323, pp. 482-496. He was referring to an article of mine in the *New Statesman* called "New Minds for the New World" (Sept. 6, 1956). As I was still in Government employment at the time, my friends in Whitehall preferred me not to sign this article; but the authorship was an open secret.

5. R. F. Harrod, *The Prof* (Macmillan, London, 1959), pp. 15, 107. Sir Roy Harrod's book is a biographical memoir of Linde-

mann. Harrod knew his subject intimately, but would not claim to understand Lindemann's scientific life.

6. T. P., autobiography, MS, p. 52.

7. *Ibid.*, p. 66.

8. *Ibid.*, p. 122.

9. *Ibid.*, p. 124.

10. F. W. Aston spent years of his life developing the mass spectrograph, and Wilson years of his on the cloud chamber: both were Nobel prize winners. Sir Thomas Merton is a distinguished spectroscopist and, incidentally, a distinguished art connoisseur and collector.

11. He had become second-in-command to Bertram Hopkinson, who was in effect head of aircraft research. Hopkinson, the most eminent academic engineer of his generation, was killed piloting his own aircraft in 1918: he, more than anyone, taught Tizard what military science meant.

12. One can, of course, make psychological guesses. It would be fairly easy to make plausible guesses about both Roosevelt-Hopkins and Churchill-Lindemann.

13. Rowe played an important part, easy to underestimate because the whole of it was secret, in the scientific war, 1935–45. He is best known as the superintendent of the Telecommunication Research Establishment, the most brilliant and successful of the English wartime research establishments.

14. It is worth noticing that Wimperis, who was a peace-loving, sweet-natured man, ill-at-ease among violent disputes, both got the committee going and selected Tizard.

15. In 1948.

16. Lord Swinton's part in these preparations, like Rowe's, though for different reasons, has been constantly underestimated.

17. At this time Sir Maurice, later Lord, Hankey. One of the

great invisible influences in English affairs, particularly military affairs, for a generation. His part has not yet been properly described.

18. Later head of the Civil Service and now Lord Bridges.

19. Cf. P. M. S. Blackett, "Tizard and the Science of War," *Nature* 185, 647–653 (1960).

20. "Operations research" in the United States. But the English started it, and I much prefer the English name. In the 1914–18 war, A. V. Hill's scientists were testing antiaircraft gunnery and were carrying out what we should later have called operational research.

21. P. M. S. Blackett, "Operational Research," *Brassey's Annual* (1953), 88–106.

22. Cf. W. S. Churchill, *The Second World War* (Cassell, London, 1948), vol. 1, pp. 399–401, 593–594.

23. Not 1937 as stated in Churchill, p. 120. There are other inaccuracies in the chapter ("Problems of Sea and Air, 1935–1939," pp. 115–128).

24. This is Blackett's account. Rowe is inclined to think, without being certain, that this critical quarrel took place before a meeting. It may easily have happened that, since a row was expected, the secretaries were told not to come in at the beginning.

25. Except Hankey. That most discreet of men, who never let slip a secret in his life, thought this was the time to do so.

26. T. P., diary, May 10, 11, 1940. R. A. E. is the Royal Aircraft Establishment; A. I. is air interception; C. H. is the first-stage chain interception; G. L. is the training of searchlights in combination with anti-aircraft guns.

27. T. P., diary, June 4, 1940.

28. *Ibid.,* June 7, 1940. S. of S. is Secretary of State.

29. *Ibid.,* June 21, 1940. C. A. S. is Chief of Air Staff.

30. *Ibid.,* July 30, 1940. M. A. P. is Ministry of Aircraft Production.

31. *Ibid.,* August 1, 1940.

32. The magnetron is a device for producing beams of high-frequency radio waves. All the advances in radar after 1940 depended upon it.

33. Blackett, "Tizard and the Science of War," loc. cit.

34. Harrod, *The Prof,* pp. 74–75, had clearly not been told the nature of the argument, either in this matter or (pp. 176–178) in the prewar quarrel.

35. The controversy would have had to be submitted with a large amount of factual background, such as the way in which aircraft are actually operated in practice. It was precisely in the misuse of this factual background that Lindemann's statistics went wrong.

36. That is, of the independent scientific members. Wimperis and Rowe were also on Tizard's side.

37. *The New Men* (Macmillan, London, 1954), pp. 278–279.

38. An interesting field of investigation would be the British Broadcasting Corporation, which, despite the Kafka-like impression it makes on outsiders, must provide some textbook examples of hierarchical politics.

39. There is a story that a small deputation of Fellows of the Royal Society called on Churchill and said that they distrusted Lindemann's scientific judgment. It would have made a pleasing scene; but I have, with regret, satisfied myself that the story is not true.

40. Some examples of these political processes enter into my novels, cf. *The Masters, The New Men, Homecomings, The Affair.*

41. I owe this remark, which seems to me truer the more I think of it, to Pamela Hansford Johnson.

42. *The Two Cultures and the Scientific Revolution* (Cambridge Uni-

versity Press, Cambridge, 1959). This was the Rede Lecture for 1959.

43. I am using "gadget" to mean any practical device, from an egg beater to a hydrogen bomb. The kind of mind which is fascinated by the one is likely to be fascinated by the other.

44. Rowe, who saw more of the English scientific choices between 1935 and 1945 than any single man, is inclined to think that, of all the scientists he met, Lindemann had the worst judgment. Judgment, that is, of science applied to war. (Letter to C. P. S., Aug. 3, 1960.)

45. They did not feel this, of course, about Tizard himself.

46. That is, those parts of science which are directly applicable to war.

47. It is, of course, also true that the feeling of society is deeply involved in military and medical science, and lays great stress upon them. If a similar stress were laid on the problems of transport, we might get scientific solutions quite quickly.

48. Cf. Elting E. Morison's biography of Stimson, *Turmoil and Tradition* (Houghton Mifflin, Boston, 1960), pp. 613–643.

49. *Saga of Burnt Njal,* chapter 113. "Foresight" in modern translations sometimes appears as "prescience."

APPENDIX

1. Charles Webster and Noble Frankland, *The Strategic Air Offensive Against Germany* 1939–1945 (H. M. Stationery Office, London, 1961), vol. 1, p. 324.

2. Cf. P. M. S. Blackett, "Tizard and the Science of War," *Nature* 185, 647–653 (1960).

3. Cf. Sir Charles Webster, *Sunday Times,* 9 April 1961; *Times Literary Supplement,* 14 April 1961.

4. The Earl of Birkenhead, *The Professor and the Prime Minister* (Houghton Mifflin, Boston, 1962), p. 202. (This book was

published in England under the title *The Prof in Two Worlds;* page references are to the American edition.)

5. Cf. Sir Frederick Brundrett, *Sunday Telegraph,* 29 October 1961. See also Professor R. V. Jones (who was a close friend of Lindemann himself), *The Listener,* 30 November 1961.

6. Birkenhead, p. 207.

7. *Sunday Times,* 5 November 1961.

8. Cf. Derek Jackson's account, Birkenhead, pp. 246-247.

9. J. T. Randall's letter to C. P. S., 9 January 1962.

10. *Science and Government,* p. 45, original edition.

11. Cf. A. J. P. Taylor, *The Observer,* 9 April 1961.

12. Birkenhead, pp. 203-204.

13. *The Spectator,* 1 December 1961.

14. Birkenhead, p. 3.

15. *Ibid.,* p. 5.

16. *Ibid.,* pp. 46, 130.

17. Cf. R. F. Harrod, *The Prof* (Macmillan, London, 1959), pp. 29-30.

18. Birkenhead, p. 360.

19. Birkenhead and Harrod.

20. Birkenhead, p. 11.

21. Cf. *Science and Government,* p. 15.

22. Birkenhead, pp. 205-207.

23. *Ibid.,* p. 209.

24. Cf. Sir Charles Webster, *Sunday Times,* 9 April 1961.

25. Blackett, "Tizard and the Science of War," *Nature* 185, 647-653 (1960), and *Scientific American,* April 1961.

26. *Time & Tide,* 6 April 1961.

27. In a letter to C. P. S.

28. A. V. Hill, *The Ethical Dilemma of Science* (Rockefeller Institute Press, New York, 1960), pp. 265-271.

29. *Daily Telegraph,* 7 April 1961, and in conversation with C. P. S.

30. *Time & Tide,* 6 April 1961.

31. Birkenhead, p. 217.

32. Webster and Frankland, vol. 1, pp. 331–336.

33. S. W. Roskill, *The War at Sea,* 1939–1945 (H. M. Stationery Office, London, 1954–1961), vol. 2, pp. 83–85.

34. Blackett, *Scientific American,* April 1961.

35. Roskill, vol. 2, p. 371.

36. Birkenhead, p. 210.

37. *Ibid.,* pp. 252ff.

38. Webster and Frankland, vol. 1, pp. 400–401.

39. *Ibid.,* vol. 2, pp. 142–145.

40. Birkenhead, p. 220.

41. Webster and Frankland, vol. 1, p. 371.